The Battery Builder's Guide:

How To Build, Rebuild and Recondition Lead-Acid Batteries

by Phillip Hurley

copyright ©2008 Phillip Hurley
all rights reserved

illustrations and photography
copyright ©2008 Good Idea Creative Services
all rights reserved

ISBN-13: 978-0-9837847-5-3

Wheelock Mountain Publications
is an imprint of
Good Idea Creative Services
Wheelock VT
USA

Copyright ©2008 by Phillip Hurley and Good Idea Creative Services

First print edition 2013

Notice of Rights

All rights reserved. No part of this book may be reproduced or transmitted in any form or by any means, electronic, mechanical, photocopying, recording or otherwise without prior written permission of the publisher. To request permission to use any parts of this book, please contact Good Idea Creative Services, permission@goodideacreative.com.

Wheelock Mountain Publications is an imprint of:
  Good Idea Creative Services
  324 Minister Hill Road
  Wheelock VT 05851 USA

ISBN 978-0-9837847-5-3

Library of Congress Control Number: 2012956489

Library of Congress subject headings:
  Lead-acid batteries--Design and construction
  Lead-acid batteries--Recycling
  Lead-acid batteries--Maintenance and repair
  Solar batteries--Handbooks, manuals, etc.
  Solar batteries--Amateur's manuals

Disclaimer and Warning

The reader of this book assumes complete personal responsibility for the use or misuse of the information contained in this book. The information in this book may not conform to the reader's local safety standards. It is the reader's responsibility to adjust this material to conform to all applicable safety standards after conferring with knowledgeable experts in regard to the application of any of the material given in this book. The publisher and author assume no liability for the use of the material in this book as it is for informational purposes only.

Table of Contents

An INTRODUCTION to WORKING WITH BATTERIES

Safety Recommendations for Battery Builders 2
Protective equipment 3
Rehearse for safety 4
Safety for lead burning 6
Mixing chemicals safely 6

An Introduction to Lead Acid Batteries 8
The chemistry of lead acid batteries ... 10
Charging 11
Types of lead acid batteries ... 11
Hybrid batteries - the ultrabattery 13
Plante and Fauré 14
Life cycle of pasted vs. solid plates 15
Alloys vs. solid lead for battery plates 16

Battery Components 18
Battery case 18
Plates 19
Straps 19
Terminal posts and intercell connectors 20

DESIGNING BATTERIES

Basic Battery Design 24
System voltage 24
Packaging the battery 25
Amp-hour capacity 25
Choosing plate type 25
Electrolyte and specific gravity 26
Battery cases 27

Estimating Cell and Battery Output 31
Cell Voltage 31
Amp-hour capacity 31
Design parameters and characteristics of Plante plates 31
Estimating output by weight of sheet lead 32
Design parameters and characteristics of pasted plates 33
Calculate amp-hour capacity by volume 34
Calculate weight of active material 35
Calculate amp-hour capacity by weight 35

Electrolyte 36
Measuring specific gravity 36
Desirable range of sulfuric acid concentration 36
Electrolyte preparation 38
Mixing electrolyte solution 40
Determining electrolyte volume 41
Calculate the specific gravity range 41
Calculate the total weight of electrolyte per cell 42
Calculate the volume of electrolyte per cell 44

REBUILDING and RECYCLING BATTERIES

Reconditioning and Rebuilding Batteries 48
Reconditioning vs. rebuilding 48
Assessing discarded batteries 48

iv

Table of Contents

Reconditioning 49
Battery reconditioning by charging 50
Water treatment 52
Battery malfunctions 53

Recycling Battery Parts 56
What battery parts can be reused
    or recycled? 57
Recycling lead for battery building 58
Handling used batteries 60
Draining the electrolyte 60
Opening the battery case 64
The mechanical approach 64
Applying heat 65
Assess the condition of the lid 65
Removing the battery elements 65
Separating the groups of plates 67
Cleaning the case 70
Repairing case covers 70
Processing negative plates
    for reuse 71

LEAD CASTING

Tools for Melting and
Casting Lead 76
Stoves for melting lead 76
Crucibles 77
Heat and splash protection 79
Other miscellaneous useful tools 79

Lead for Foundry Work 80
Melting new lead 80
Melting salvaged lead 81

Making Molds for Lead Casting 82
Patterns 82
Making forms 84

Mold platforms 86
Release agents 88
Plaster of Paris 89
Mixing plaster 89
Pouring the mold 90
Removing the pattern and form 91
Cleaning and curing the mold 92

Pouring the Lead 94
Practice pouring 94
Using flux 94
Casting straps without slots 95
Casting straps with slots 96
Aligning the mold 97
Melt and pour 99

Finishing Castings 101

Casting Connecting Rods
and Seals 104
Connecting rods 104
Casting pocket nuts 105
Casting sploot seals 108

Casting Plates 111
Methods for casting plates 111
Making a plate mold 112
Release agent 114
Melting the lead 114
Measuring lead for the melt 115
Level the stove 115
Simple plate mold 118
Finishing the casting 119
Casting plates without a mold 120

Working with Sheet Lead

Tools for Working Sheet Lead 126
- Smoothing sheet lead...................... 126
- Lead cutting tools........................... 127
- Making the cuts............................. 128

Making Plates from Lead Sheet....130
- Smoothing the lead........................ 130
- Finishing the plates........................ 139

Adding Plate Texture..................140
- Surface treatment for Plante plates.... 140

Preparing the Plates for the Burning Rack 144

Pasted Plates

Making Plate Grids from Sheet Lead 146

Mixing and Applying Paste...........147
- Using a binder............................... 147
- Mixing in the electrolyte 148
- Paste formula and preparation 149
- Safety procedures 150
- Set up the work area...................... 152
- Pasting container........................... 152
- Prepare the grid plate..................... 153
- Measuring the ingredients 153
- Prepare the dry ingredients............. 154
- Adding wet ingredients.................... 155
- Pasting the plates.......................... 156
- Curing and drying the grids 157

Lead Burning

Plate Burning Rack 160
- Design and function....................... 160
- The base of the rack 161
- Side slider rims 162
- Combs 162
- Leveling feet 164
- Uprights 165
- Tools... 166
- Materials 166

Tools for Lead Burning.................167
- Torches 167
- Putty for lead burning 168
- Preparing surfaces for burning.......... 168
- Cleaning the work pieces................. 169
- Lead wire..................................... 170
- Post molds 171

Burning Lugs onto Plates 172
- Burning lugs 172
- Laying in a melt............................. 176

Burning Plates into Groups 178
- Set up components in the rack.......... 178
- Prepare the components.................. 180
- Set up heat shield cloth................... 181
- Set up the containment dam 181
- Final check 184
- The burn...................................... 185

Battery Assembly

Plate Separators 190
- Materials used for separators........... 190
- Separator shape 190
- Ribbing 190
- Mesh 191
- Making sleeve separators 192

Table of Contents

Battery Assembly 197
- Assemble the elements 198
- Fit and install the elements 199
- Install the cover 203
- Add the electrolyte 205

Charging and Forming

Plate Forming 208
- Forming and charging basics 208
- Designated vs. actual voltage 208
- Forming for Plante plates 208
- Forming cycles 208
- Discharging 209
- Initial charging for pasted plates 211

Equipment for Forming and Charging 212
- Charging from the grid 212
- Manual chargers 212
- Smart chargers 213
- Charging multiple batteries 213
- Chargers for either grid or solar 214
- Solar chargers 215

Charging and Forming Procedure 218
- Sequence for charging 218
- Discharging sequence for forming Plante type plates 220
- Next cycle for forming 221
- Final charge 221
- Gassing 222
- Electrolyte levels 222
- Wiring and Connectors 224
- Formulas for working with batteries 226
- Tests, performance characteristics and statistics 227

Appendix
- Information Resources 232
- Materials and Tools 235
- Other Titles of Interest 240

The Battery Builders Guide

An INTRODUCTION to WORKING WITH BATTERIES

Safety Recommendations for Battery Builders

The materials discussed in this book are dangerous. Lead is toxic, and sulfuric acid is extremely corrosive. To work with lead metal, lead oxides, and sulfuric acid you must know and implement protective measures for the safe and appropriate handling, processing, transfer, transport, and disposal of these toxic and caustic materials.

Sulfuric acid, if splashed in the eyes, will cause blindness; and, if it contacts the skin, will cause severe burns. You absolutely should never work with sulfuric acid and lead without appropriate eye, face, respiratory and skin protection. Lead is toxic to everyone but is particularly dangerous for pregnant women and the unborn fetus, and small children, so in addition to safety concerns for yourself, you must consider the safety of those close to you, the general public, and the environment.

- You will need adequate workspace and ventilation.
- Children and pets should never be allowed access to your work or storage areas.
- Acquire and thoroughly read the MSDS (Material Safety Data Sheet) for every substance you intend to use.
- When working with lead metal, lead oxides and sulfuric acid you must have appropriate body cover and respirators to protect yourself from the ill effects of these substances.
- Learn, understand and apply the rules and regulations that your AHJ requires you to follow.

AHJ stands for Authority Having Jurisdiction. This can be any number of official offices at the federal, state, local level which have authority to define and require compliance at a given place. Rules, regulations, and approved practices vary widely from one jurisdiction to another across

Safety Recommendations

the globe. It is not possible in this text to present the exact requirements in force for your locale. It is up to you to research, access, and apply the appropriate and required standards for where you reside.

That being said, there are common sense guidelines freely available on the internet that apply to working with the substances mentioned in this book. The websites of the organizations below and the MSDS sites listed in the Appendix are great resources for relevant safety information.

- **NIOSH** (National Institute for Occupational Safety and Health) — *Pocket Guide To Chemical Hazards*, and *Recommendations For Chemical Protective Clothing* are two publications available online at the NIOSH website. These are the best guides for protective devices and clothing. Also, they have a useful respirator fact sheet.

- **CCOHS** (Canadian Center For Occupational Health And Safety) — useful information about chemical protective clothing and glove selection.

- **OSHA** (Occupational Safety And Health Administration) — lead battery manufacture.

- **EPA** (U.S. Environmental Protection Agency) — treatment, storage, and disposal of hazardous waste.

- **DOT** (U.S. Dept. of Transportation) — safe transport of hazardous materials.

Protective equipment

There are many distributors of protective clothing, face masks, and respirators listed on the web who sell the appropriate gear for working with batteries. Most industrial suppliers such as McMaster-Carr and others listed in the resource section of this book will be very helpful in selecting just what you will need if you tell them what substances you will be working with.

The Battery Builders Guide

When working with sulfuric acid I use a full PVC or Viton® suit which consists of pants, top with head cover, and a face shield with two pairs of natural rubber, Viton® or other type of acid resistant gloves. I use two pair so that if the skin of one breaks I will still be protected by the other layer.

I use a full face mask and a separate respirator. There are full face masks with built-in respirators which are nice.

Be sure to have an eye wash kit, eye wash station available and ready to go and get to know how to use it before you work with acid.

Use the buddy system when working with acid. Have someone else there who is also suited up and knows emergency procedures in case something goes wrong. This can be very useful in case of an accident.

Rehearse for safety

When doing any work with toxic materials and caustics, I always do a "dry run" to figure out the best position for various items, and the movements that will be the safest and take the least time. This also helps to ensure that all the materials and equipment needed are at the ready to complete each operation that you plan to do. Also, before using any

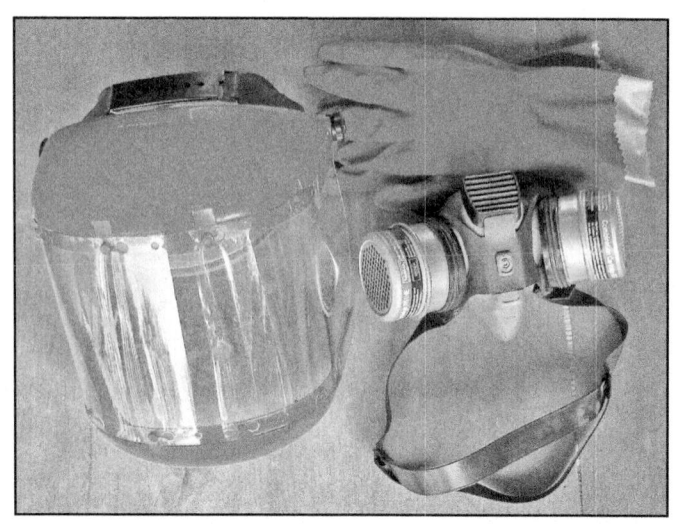

Face mask, two pairs of gloves and respirator

Safety Recommendations

equipment for final work, be sure to acquaint yourself with each tool (stove, torch, etc.) and practice using it before beginning work.

Another vitally important safety practice is to keep all unauthorized personnel away from your work area. If there are pets or children around, you must make absolutely sure that they cannot access your workspace. Sulfuric acid, for instance, looks just like water to children and pets, and uninformed adults, for that matter, if it is not adequately labeled. You do not want to be responsible for seriously injuring someone for life.

Keep plenty of baking soda on hand when working with acid

Make sure your work area is clear and unobstructed. Wherever you work, you must have adequate ventilation so that you are not breathing noxious fumes, dusts, and vapors.

When working with acid, always have a 5 gallon bucket of water with plenty of dissolved baking soda in it to neutralize the acid from your gloves and other items that need neutralization. I also keep a large quantity of dry baking soda available to neutralize possible spills or drips.

Decontamination soap

When working with lead, use a decontamination soap such as D-Lead to wash yourself afterwards. Wear gloves whenever possible when handling lead, and avoid skin contact.

Safety for lead burning

- When working with torches, gas stoves, soldering irons or wood burning irons, be sure that work surfaces are fire resistant.

- Wear heat resistant gloves for handling hot melt pots for pouring.

- Always wear some type of eye protection when working with molten lead.

- Wear clothes that will protect you from any splashes from molten lead.

- Do not work with molten lead when it is raining or snowing if your work area is at all exposed to the weather; and never work with a moist or wet mold. Any moisture in contact with molten lead can cause an explosion of hot flying lead. Antimony, arsenic and calcium are often alloyed into lead. When the dross of these alloys are mixed and moisture is present, the highly toxic gases stibine and arsine can be produced.

Mixing chemicals safely

Never pour water into sulfuric acid as it will react violently. Always pour the acid into the water, or the most concentrated solution (stronger acid) into the less concentrated solution (weaker acid) when mixing electrolyte. Plexiglas® shields can be employed with glove inserts to avoid acid splashes.

When working with lead oxides it is a good idea to work with a glove box

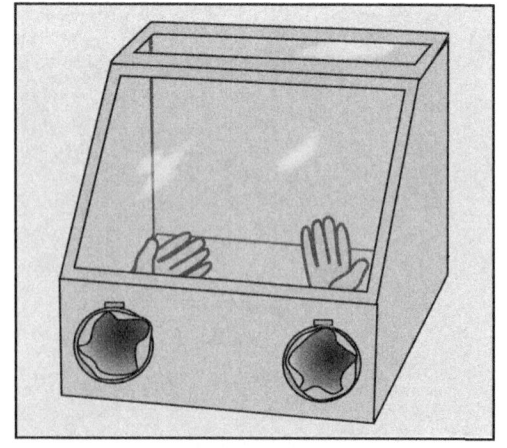

Glove box

Safety Recommendations

or fume hood. They are easy to build and help contain airborne particles during mixing.

Don't forget that as human beings we all make mistakes sooner or later, or circumstances and events will transpire to upset our most carefully laid-out plans. Do not gamble! Always wear protection and follow practices that will keep you and others out of harm's way.

Please note that any information and suggestions given in this text are not a substitute for the safety, health, and environmental practices required in your locality by your AHJ.

An Introduction to Lead Acid Batteries

Lead acid batteries are energy storage devices composed of two or more individual electrochemical cells connected either in series or parallel.

The average lead acid cell has a nominal voltage of 2.2 volts when fully charged and is usually referred to as a 2 volt cell. Series connected cells will add the voltage of each cell to provide a higher voltage. Parallel connected cells will add the current (amperage) of each cell to provide a higher current (amperage). Most available commercial batteries are comprised of series connected cells.

Each cell within a battery contains one positive and one negative electrode. Each positive and negative electrode can consist of one plate each or many positive plates and many negative plates interleaved.

Left, 3 cells connected in series. Voltage is added, so the 3 cells make a 6 volt 1 amp battery.

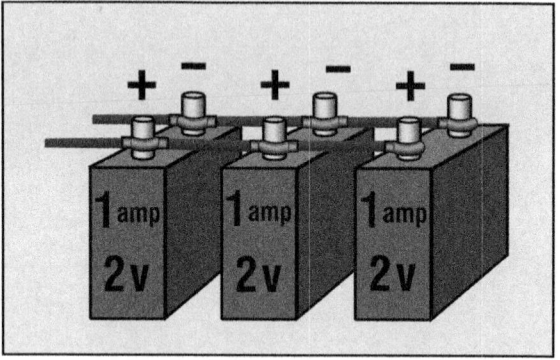

Right, 3 cells connected in parallel. Amperage is added, so the 3 cells make a 2 volt 3 amp battery.

If an electrode has more than one plate, the plates are connected together with a conductive strap, and called a group.

Introduction to Lead Acid Batteries

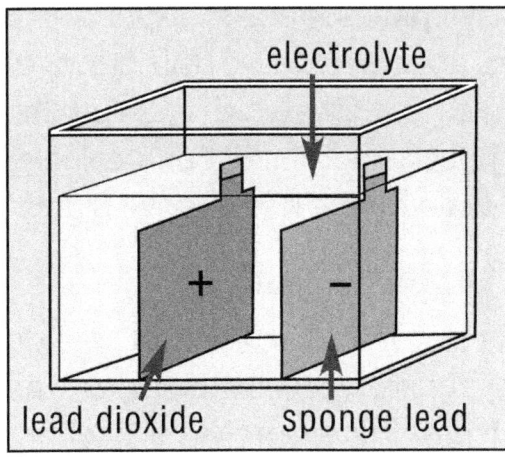

Simple battery cell with one plate each for the positive and negative electrodes

Although the voltage is limited to a little over 2 volts per cell in a lead acid battery, the current is only limited by the area available on the plates for the active materials engaged in the electrochemical process. The larger the surface area, the more current can be available from the cell and battery. Rather than having two or more very large plates it is convenient and more practical to use a parallel configuration of many smaller size plates to deliver more current.

Most battery cells are thus composed of a group of negative plates and a group of positive plates which are interleaved and insulated electrically from each other by a non-conductive separator.

This arrangement is called an element. Each element consists of an odd number of plates.

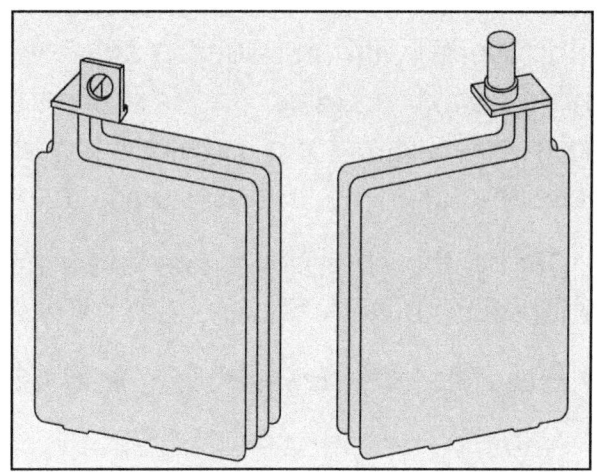

Left, a negative electrode group, and right, a positive electrode group.

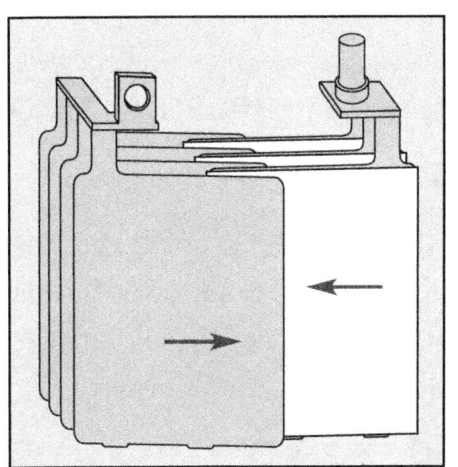

An element, consisting of one positive and one negative group, interleaved with each other. Each positive plate is enveloped in a non-conductive separator.

9

Because the active material on the positive plates tends to expand and contract considerably, the outside plates are always negative. This arrangement works both sides of the positive plates equally, which reduces the chance of the positive plates warping and buckling in use. So, a cell element will always have one more negative plate than it has positive plates.

The chemistry of lead acid batteries

In a charged battery, the negative electrode (anode) consists of sponge lead and the positive electrode (cathode) consists of lead dioxide. The sponge lead ($PbSO_4$), lead dioxide (PbO_2) and the electrolyte (HSO_4) are referred to as active materials. They actively participate in the electro-chemical charge and discharge process.

Each positive and negative plate within a cell is insulated from other plates with a non-conductive porous separator. The plates are immersed and covered with an aqueous solution of sulfuric acid and distilled water. This solution is called an electrolyte. The electrolyte participates in the chemical reaction and provides ion transport between the negative and positive plates (electrodes) during the charge and discharge cycle.

During the charging process, electrical energy is applied to the cells where it is converted into chemical energy:

Positive electrode — $PbSO_4 + 2H_2O \rightarrow PbO_2 + 3H^+ + HSO_4^- + 2e^-$

Negative electrode — $PbSO_4 + H^+ + 2e^- \rightarrow Pb + HSO_4^-$

This chemical energy is then reconverted into electrical energy during the discharge process:

Positive electrode — $PbO_2 + 3H^+ + HSO_4^- + 2e^- \rightarrow PbSO_4 + 2H_2O$

Negative electrode — $Pb + HSO_4^- \rightarrow PbSO_4 + H^- + 2e^-$

During the discharge process, lead dioxide on the positive plate and the sponge lead on the negative plate react with the sulfuric acid to form lead sulphate, water and electrical energy. The specific gravity of the electrolyte is reduced during discharge due to loss of sulfuric acid in solution.

$$PbO_2 + Pb + 2H_2SO_4 \rightarrow 2PbSO_4 + 2H_2O$$

During the charging process, the lead sulphate on the plates is electrochemically converted to sponge lead on the negative plate; and lead dioxide on the positive plate and sulphate ions go back into the electrolyte to form sulfuric acid. The specific gravity of the electrolyte rises during charge due to re-formation of sulfuric acid in the electrolyte solution.

$$2PbSO_4 + 2H_2O \rightarrow PbO_2 + Pb + 2H_2SO_4$$

Charging

Lead acid batteries can be charged from any voltage regulated DC (direct current) source, and when charged can provide DC electricity for a variety of mobile and stationary applications and appliances. The DC electricity from batteries can also be processed through an inverter which changes it to AC (alternating current) which can then power most common household AC appliances.

Types of lead acid batteries

The most commonly available lead acid batteries are 6 volt batteries comprised of three 2 volt cells connected in series, or 12 volt batteries comprised of six 2 volt cells connected in series. Individual 2 volt cells are used for photovoltaic and industrial use, but they are not batteries per se as they are only one cell. Two of these individually packaged cells become a battery when they are connected to each other either in series or parallel.

Lead acid batteries come in many shapes and sizes, and different current and voltage ratings. Some are sealed and do not require water maintenance:

- **SLA (sealed lead acid)** are also known as low maintenance, recombinant, gel cell, or AGM, absorbent glass matt, or sealed VRLA (valve regulated lead acid) batteries.

Three different types of SLA (sealed lead acid) batteries and cells

Some are not sealed and require occasional water replenishment:

- **FLA (flooded lead acid)** are also known as vented cell or wet cell batteries.

Both sealed and flooded batteries have vent or valve gas escapes for over-pressure release of gases, but the sealed type do not allow replacement of the lost water. Sealed batteries are low maintenance, but they require close monitoring of charging as overcharging can destroy them. The flooded non-sealed types are more forgiving: if they are overcharged they can be replenished with distilled water.

Both sealed and flooded batteries can be classified as SLI (starting, lighting, ignition), marine, or deep cycle:

- **Deep cycle/industrial/traction batteries** are designed for constant and continual use, with greater depth of discharge. The lead plates used in these types are thicker than SLI types. They are mainly used for alternative energy systems, golf carts and fork lifts, where a continuous, more constant power supply is required.

Introduction to Lead Acid Batteries

- **Marine batteries** fall in between SLI and deep cycle and have medium thick plates relative to the other two types. They provide a greater allowable depth of discharge and, as the name implies, they suit the marine industry well for water craft applications.

- **SLI** are mostly used for automotive electrical systems and are designed for intermittent use where high currents are needed for short periods of time. The common automobile starting battery is of this type and has many thin lead plates. These give a greater surface area for active material in a smaller space. This allows more current output for short intervals, which is desirable for many automotive applications.

Hybrid batteries — the ultrabattery

Another category of lead acid battery is what are called ultrabatteries or superbatteries. These are hybrid batteries: a lead acid battery with a super-capacitor integrated into each cell. They are a new and very promising innovation, especially for hybrid and electric vehicles that use regenerative braking, where a battery must maintain a partial charge at 60% to 90%. Combining a lead acid battery with a supercapacitor permits rapid repeat charging and release of energy, so the battery is far more efficient at using and releasing the energy available to it. In addition to use in electric and hybrid vehicles, ultrabatteries could be handy in a variety of applications, including stationary solar, when higher surge currents are needed to start motors or power other high momentary current demands.

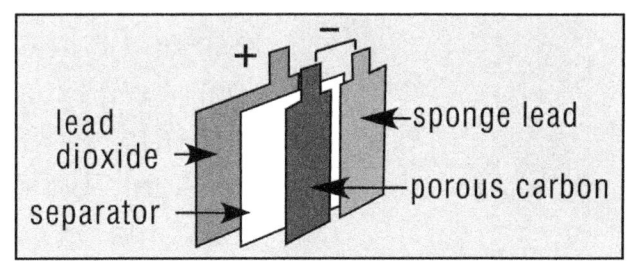
Ultrabattery

The porous carbon negative plate, which is added to each negative lead plate, has a very large active surface area. It can be made of carbon aerogels, activated carbon, or nano-carbon.

Plante and Fauré

Gaston Plante invented the lead acid battery in 1859. His first battery cell consisted of two sheets of lead that were about 60 centimeters (23.6") long by 20 centimeters (7.8") wide and one centimeter (0.039") thick. He coiled up these sheets with a piece of felt between them for insulation and immersed

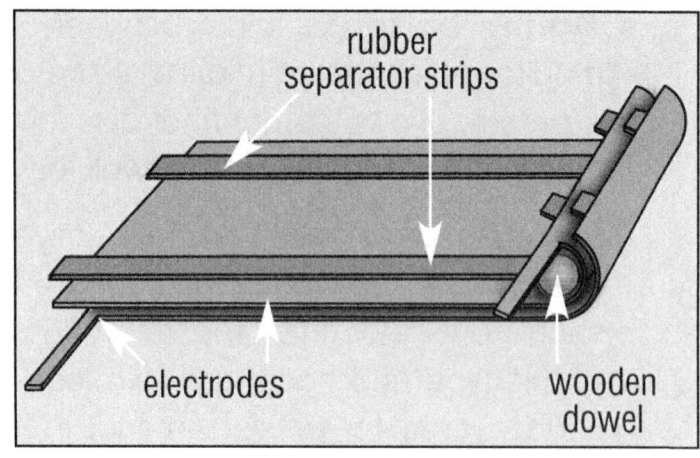

Above and below left, one version of a Plante spiral battery

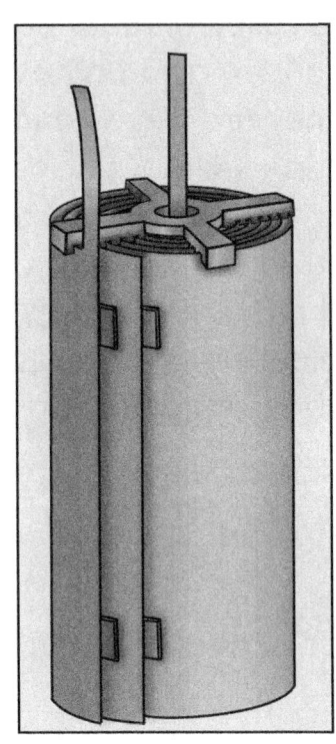

them in a jar with about an eleven percent solution (sp gr 1.070) of sulfuric acid electrolyte. He then formed the cells using a series of charge and discharge cycles while reversing the charge polarity at each successive charge. This process, called formation, alters the characteristics of the plate surfaces by spongifying the lead on each plate, which enhances the performance of the cell. The porous sponge plates of Plante's experimental cell produced about 7.25 amp-hours per pound of lead and had an efficiency of about 72%. Plante continued experimenting and improving on his lead acid cell for the next twenty years. Although the Plante batteries worked very well, they were not viable as a commercial offering because the process for forming the plates took a long time.

In 1881 Émile Alphonse Fauré developed a different process for plate preparation and drastically reduced the time it took to make plates. Rather than forming the solid plates by the tedious process of charging and de-charging through many cycles over many days, Fauré filled grids with the appropriate chemical compounds so that the batteries could be put into service after one initial charge. This made the manufacture of lead acid batteries an economical enterprise, as they could be mass produced at a fast rate.

Life cycle of pasted vs. solid plates

Most lead acid batteries today have Fauré plates or some variation of the Fauré plate principal, such as a tubular plate. Planté plates, however, are still used and preferred for heavy duty applications where long duty cycle is a concern. They are more expensive than Fauré plate batteries or derivatives, because they generally use thicker lead plates and take longer to form. They also are usually heavier than Fauré plate batteries because they require more plate surface area to provide the same density of current that smaller, thinner pasted Fauré plates provide. Fauré plates, however, are prone to shedding their active material, which reduces their efficiency over time. Planté types keep on producing active material from the abundant available lead in their thicker solid plates, so they have greater longevity.

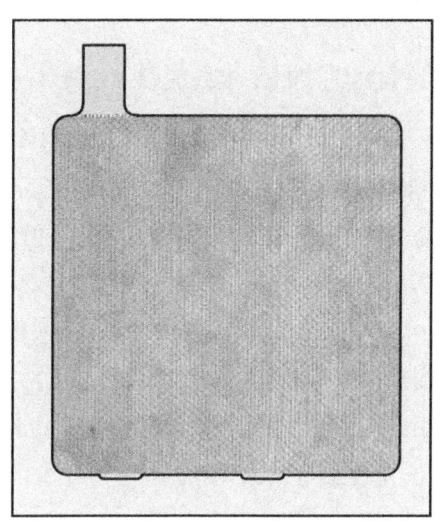

A solid plate, with scoring to increase surface area.

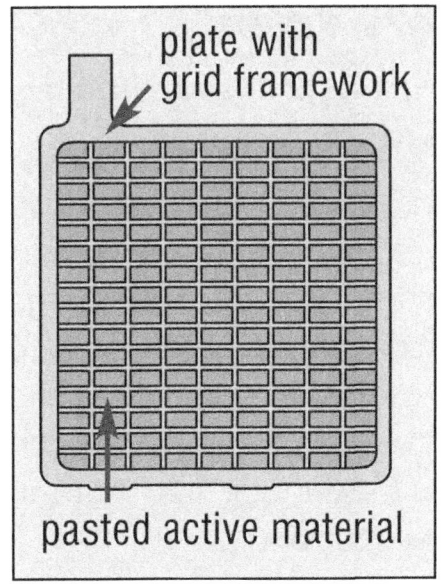

A pasted Fauré plate

A good compromise to give Fauré type batteries a longer life is a tubular plate which contains the active material in porous plate tubes. This greatly reduces active material loss during the lifetime of the battery. The tubes are fabricated from a porous non-conductive material like ceramic or plastic, with a central conductive rod. The spaces between the rod and the inner walls of the tube are filled with active material. The tubular type Fauré plate is a popular offering in the deep cycle lead acid battery market today. Many deep cycle batteries consist of positive tubular plates and negative pasted grids.

Alloys vs. solid lead for battery plates

Most lead plates used in pasted lead acid batteries today are alloyed with either antimony or calcium to make casting the plates easier, and to strengthen the plate. Lead plates alloyed with antimony can be deep cycled more often than lead plates alloyed with calcium, however lead-antimony plates have higher self-discharge rates than lead calcium plates. Tin is often added to lead-calcium plates which allows them to be discharged more deeply and more often. Selenium is often added to lead-antimony plates to reduce self-discharge losses. Other dopants used to improve charge and discharge characteristics are cadmium and arsenic. Self-discharge rates can be kept at a minimum by using a lower specific gravity electrolyte, and storage at reasonably cool temperatures since batteries have higher discharge rates as the temperature increases. The self discharge rate can also be offset by trickle charging.

Basically the best choice of material and type of plates for a battery depends on the intended use of the battery.

Lead-calcium Fauré type plates are more suitable for automotive applications where short term, intermittent higher current discharges are required. Automotive batteries normally do not see deep discharge service, so calcium alloy works well in that circumstance. Because Fauré plates have the active material added to the plates, they can be made thinner. Since the plates are thinner and weigh less, a Fauré battery can have many more plates per cell. This increases the surface area that can react with the sulfuric acid and supplies a higher current.

Introduction to Lead Acid Batteries

Non-alloyed lead plates are more suitable for stationary and industrial applications which require a constant daily charge and deep discharge (slower rate, less current demand), and where the weight of the battery is not a problem.

Non-alloyed lead plate Plante type batteries have the longest life cycle. Lead antimony come in second and lead calcium come in third for longevity. Fauré type plates generally have a more limited lifetime than pure lead plates as they tend to shed active material and lose capacity over time.

Battery Components

Battery case

A battery is contained by a case that has compartments with leak proof partitions. The partitions electrically isolate each cell of the battery from the other cells. The number of compartments in a battery case determine the number of cells within a battery and thus the output voltage.

Cases are made from a variety of materials, the most common being polypropylene and polyethylene due to their excellent resistance to sulfuric acid, fair mechanical durability, and relatively inexpensive fabrication costs. Historically battery jars, containers, and cases were made from glass, wood, or hard rubber. Wood containers were covered with asphaltum or paraffin wax to provide acid resistance and to make them leakproof.

The bottom of each cell compartment in a case has ribs on which the plates rest. The raised ribs form a space at the bottom

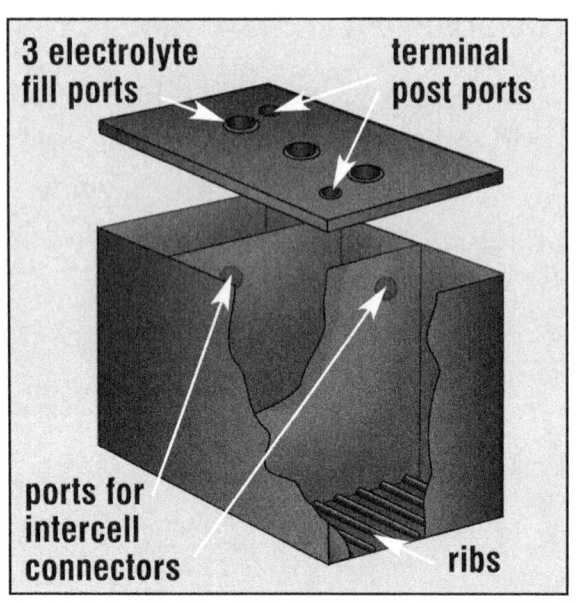

Above, case and lid for a battery with internal cell connectors. Below, case and lid for a battery with external cell connectors

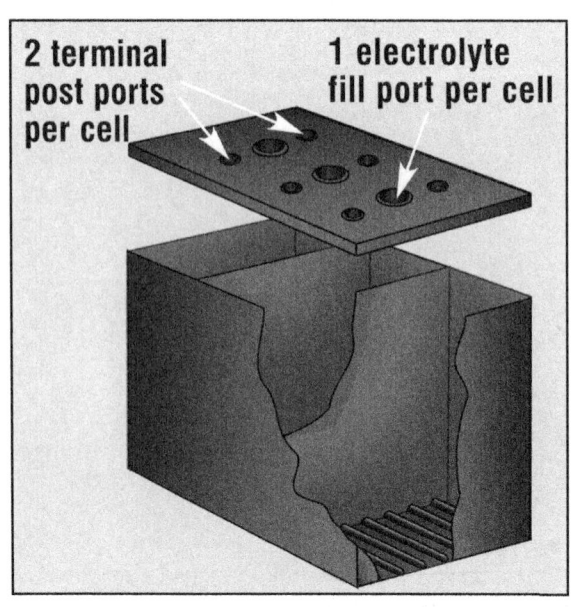

Battery Components

of the cell for sediment falling from the plates and keeps the sediment from shorting out the plates.

Each case has a lid with ports to access the electrolyte in each cell, and ports to accommodate the terminal posts. The electrolyte fill ports have caps that can be removed for checking specific gravity, checking electrolyte levels and filling with distilled water. Caps are usually either bayonet types or threaded. They have a small porthole which allows release of excessive pressure, and a baffle in the cap to prevent excessive loss of electrolyte vapor.

Plates

Plates can be either pure lead or lead alloy. They can be solid or pasted grid. Plates have an extension tab which is often called a lug for attachment to the group strap.

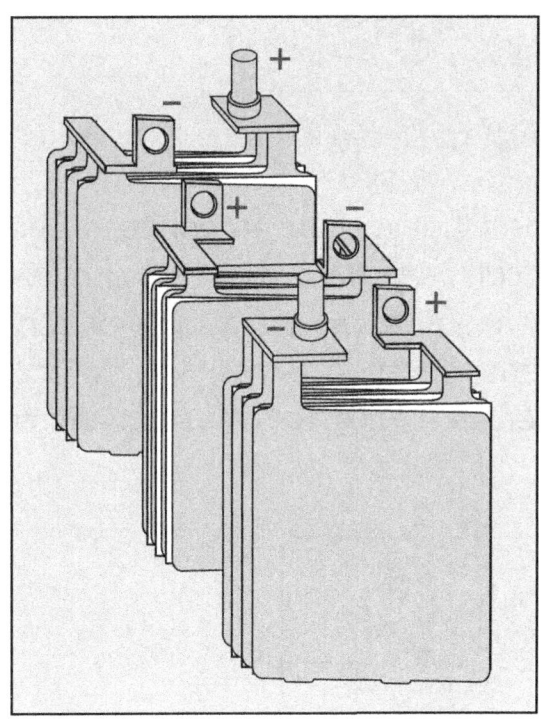

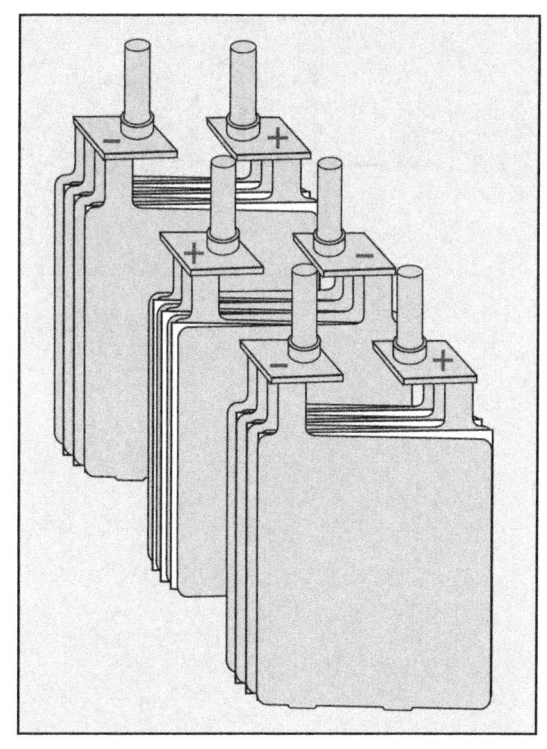

Above, 3 elements for a battery with internal cell connections; Right, 3 elements for a battery with external cell connections.

Straps

Straps or plate straps are bus bars of pure lead or lead alloy that connect all the plates in a group.

Terminal posts and intercell connectors

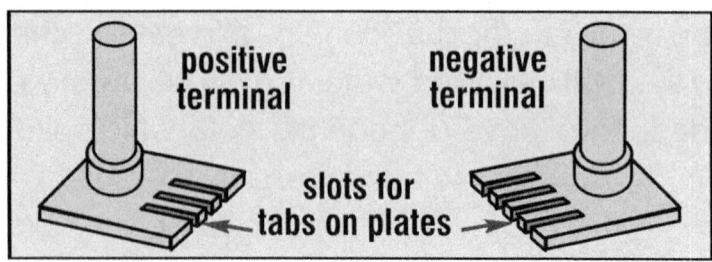

Terminal posts for external connections

Terminal posts are the visible lead posts that you see on batteries. The number of terminal posts on a battery can vary. Most consumer batteries have two: one positive and the other negative. Most individual battery cells within a case are connected in series (positive to negative) to add voltage. The cells are connected via intercell connectors which can either be external or internal.

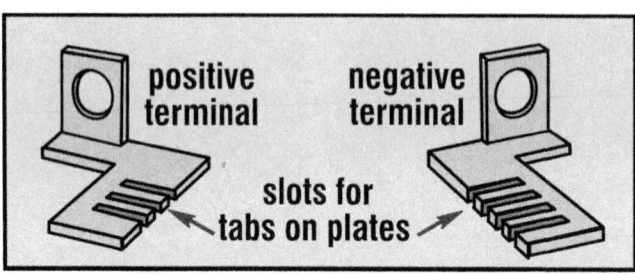

Internal intercell connections

Battery Components

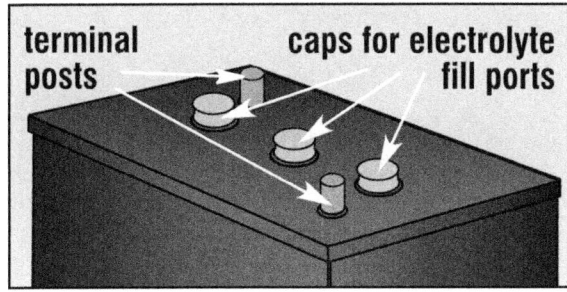

*Above, top of battery with internal intercell connections;
Below, top of battery with external intercell connections*

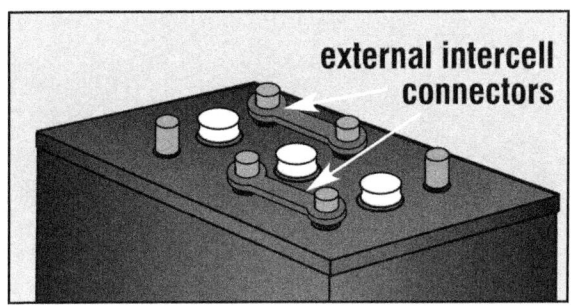

Batteries with no internal intercell connectors will have more than two external posts on the battery. A 12 volt battery, for instance, would have 12 external posts. A 6 volt battery would have six external posts.

Right, cutaway of battery with internal intercell connectors.

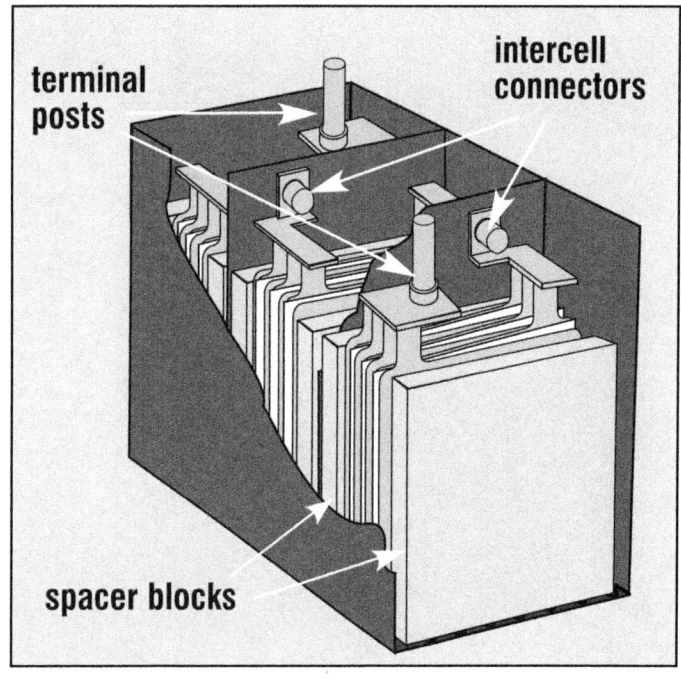

The Battery Builders Guide

DESIGNING BATTERIES

Basic Battery Design

There is no one best basic design for batteries. The best battery is one that has been designed specifically for the service the battery will perform.

The basic service requirements of a battery are referred to as the duty cycle. To determine the nature of the duty cycle and choose the appropriate components for a battery, consider the following:

- Will the service requirements be mainly intermittent (short periods of time) or continuous (longer periods of time)?
- Will the current draw during these time periods be high, medium or low?
- What are the exact power requirements of your system? Power requirements are usually expressed in watts.
- How will the battery be charged, and how long will it take to charge?
- Will the battery be stationary or be in a mobile application?
- What temperature and other environmental conditions will the battery be exposed to during operation?

Once these parameters are defined, consider what will make an individual battery within a power system fit for the duty cycle you require.

System voltage

A power system has a specific operating voltage. The individual batteries do not each have to match the whole system voltage requirements. For instance, to power a 12 volt system, instead of one 12 volt battery, you can use two 6 volt batteries connected in series, and so on, with many variations.

Packaging the battery

When the battery voltage has been determined, then consider how to package the battery. Would it be more convenient to have individual 2 volt cells in separate containers, or would it be better to have one container with the required number of cell wells? If you use one container with cell wells, will the intercell connectors be internal or external?

Amp-hour capacity

One of the most important parameters is the ampere-hour capacity of the battery. This is directly related to the active surface area available on the plates: the more surface area on the plates, and the more plates that you pack into a cell, the more amp-hours will be available.

Choosing plate type

Should you use pasted plates or solid plates, and how thick should the plates be?

Pasted plates can be used for either SLI or deep cycle applications; solid plates are usually used for deep cycle stationary power applications.

Pasted plates for SLI operation can be thinner and fewer plates used. Pasted plates shed active material during use, so when used for deep cycle operation, they need to be thicker to perform over an extended time period. Deep cycle batteries must provide more amp-hours for extended periods of use, so they need more plate surface area. Deep cycle solid plates are usually thick because they also lose active materials, but they recoup the loss through chemical action upon the newly exposed plate surface.

For practical purposes, plates should probably not be bigger than 12"x 12". Most plates for consumer type batteries are about 6"x 6" or under.

There are many methods that can be employed to make plates of either the Plante type or the grid pasted plate.

For pasted plates, recycled grids are the easiest and most economical way to go. The only limit in using recycled grids is that you have to settle for the size available and make sure you can get enough grids of the same battery type to complete your battery or batteries. This is probably a minor concern as there are usually more than enough batteries available for selective recycling. Plate grids can also be cast using recycled grids for mold patterns; or, they can be cut from sheet lead. If cut from sheet lead, the grid spaces must be cut or stamped out of the lead sheet.

Solid Plante type plates can either be cut from new purchased sheet lead, recycled sheet lead, or sheet lead that you have cast from lead, either new or recycled pieces. For solid plates the lead used should be unalloyed. Alloyed lead tends to be more resistant to the formation process. Cast lead is better to use than worked or rolled lead, however this is a minor issue.

Electrolyte and specific gravity

The specific gravity of the electrolyte affects the performance of a battery. Depending on what tasks the battery is designed to perform, the original specific gravity can range from a high of 1.300 to a low of 1.200. High specific gravity allows higher momentary current discharge since more acid is quickly available to the active surface of the plates. So, for instance, a higher specific gravity works for applications in electric vehicles where higher discharge currents are used and the batteries are charged and discharged frequently. However, the stronger acidity means that the plates will have higher rates of self-discharge and corrosion. Electrolytes with high specific gravity do not work well for batteries that are to be used for operations such as intermittent emergency standby power use, or for general stationary power applications. High specific gravity batteries need to be used frequently. They have a shorter life cycle than lower specific gravity batteries, especially with solid, non-alloyed lead Plante type plates.

Lower specific gravity electrolyte will not support the higher momentary current discharges that a higher specific gravity electrolyte can give, but it will give the battery plates a longer life as they do not corrode as fast, and they also have lower self-discharge rates. For most general applications, a specific gravity of 1.265 is a good compromise.

Battery cases

Cases can be made from any sulfuric acid resistant material. Polypropylene and polyethylene are used frequently, and glass is sometimes used in industrial and laboratory settings.

For sulfuric acid resistance, Viton® is the best and is rated to be used up to full concentration of sulfuric acid. LDPE (low density polyethylene), HDPE (high density polyethylene), PVC (polyvinyl chloride), CPVC (chlorinated polyvinyl chloride) are rated for use with up to 80% concentration of sulfuric acid. PP (polypropylene) is rated for use with up to 70% concentration of sulfuric acid, and vinylester, and polyester are rated for use with up to 50% sulfuric acid. These ratings are general and will vary according to use, and temperature. You will need to consult a professional table for acid resistance for exact, safe, and correct usage of these materials.

Cases can either be purchased new, be fabricated, or you can recycle cases. When designing cases, remember that the electrolyte should cover the top of plates by ⅜" to ½", and then add another ½" or more from the top of electrolyte to top of case cover.

Case covers need to fit well and be sealed, and have ports with caps for filling and venting. If you make your own covers, recycled caps can be used, which can save you some fabrication work. Or, you can purchase new caps like the flip-top vented caps shown in the photo, next page.

New vented flip-top caps for a recycled case lid

Case covers can be sealed with Viton® caulk and then epoxied to retain the cover to the case. If heat is used for sealing or unsealing, here are the melting points for some commonly used materials:

- Polypropylene — 320°F

- High density polyethylene — 248°F to 266°F

- Low density polyethylene — 221°F to 239°F

- Polyvinyl chloride — 413°F

If you seal covers to cases, never lift the case from around the cover top, or use any built-in battery carrier slots, or devices that are molded into the battery cover that use web type hook straps, etc. Always lift your batteries from the sides of the case walls and from the bottom. Batteries are heavy, and more than likely, no matter how well the case is glued to the cover, they will come apart if you try to lift all that weight from the cover. Avoid an accident and always lift from the case body.

All batteries that you fabricate should be labeled with manufacturers name, date, voltage, amp-hour capacity and the warning: "This battery

Basic Battery Design

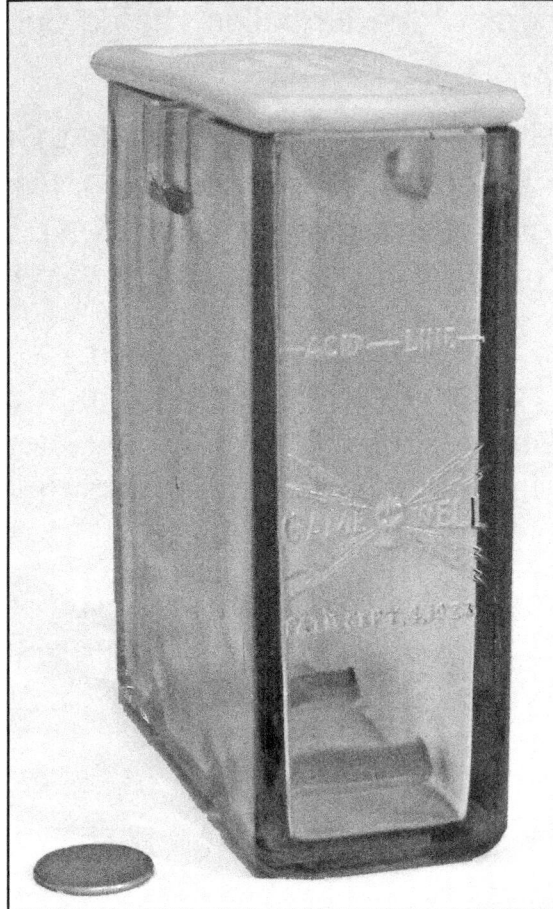

contains lead which is toxic; and sulfuric acid which is caustic and will cause severe burns."

At one time, glass was used quite frequently for cases. Many antique battery jars are still available and can still serve as good experimental or test battery jars if they are not cracked or chipped with dangerous edges. If you do have a jar that is chipped a little on the edge, cover it with paraffin (which is acid resistant). Never use a cracked jar.

Left, antique glass battery jar and lid. Below, antique resistor jar

When using glass jars, put a bead of paraffin just under and around the top inner edge of the jar to prevent electrolyte creep and condensation from forming around the top edge of the jar.

Glass jars are excellent for testing plate designs as their transparency allows you to observe some of the changes that occur during different parts of the formation charge and discharge cycle.

There are also large glass jars available that are often mistaken for battery jars, but are in fact resistor jars. These also make excellent test jars. The larger size jars are good to prep Plante plates in a dilute nitric acid bath, or use them to form Plante plates prior to insertion in the final battery case.

When using smaller glass jars or containers, always set them in a bed of sand so that they can not easily be knocked over. A kitty litter box or similar container with a few inches of sand in it to place the jar in will serve the purpose.

Please note that glass jars should never be directly placed in the sun as the heat can crack the glass. If you use glass, remember it is fragile and take appropriate safety measures to reduce risk of injury.

Estimating Cell and Battery Output

Cell Voltage

Cell voltage output in a fully charged cell is most directly related to the concentration of sulfuric acid in the electrolyte.

Amp-hour capacity

Amp-hour capacity is most directly related to the surface area of the active materials on the plates. The greater the surface area of the plates, the more current that will be available on demand for high current momentary discharge, and the more current that will be available over a longer period of time at low discharge rates.

Design parameters and characteristics of Plante plates

Plante plates are usually fabricated from unalloyed sheet lead. As noted, the greater the surface area of the plates, the more amp-hours will be available.

For Plante plates, the treatment of the surfaces of the plates is crucial to the final output of the plates. They should be scored or grooved to create more surface area for active material to form on, and to provide a lodging space for the active material. The depth and width of the scores, as well the quantity of score lines, determine the final active area available.

Score lines do not need to be more than about 1/64" deep and wide, however, they can be any width or depth according to your needs or experimental direction.

There are many ways to score and groove plates, and they all give different quantities of scoring, depth and width. For this reason, it is not possible to present a formula that will exactly determine the active area of the plates, and thus the amp-hours that each cell or battery will be able to produce. However, you can get a ballpark figure for the amp-hour capacity by calculating from the obvious area of a plate.

Estimating output by weight of sheet lead

Commonly available sheet lead is described in terms of pounds rather than by thickness of the sheet. One square foot of a given thickness of lead sheet will weigh roughly what its pound designation indicates. However, as in other trade designations, such as pipe size, the actual weight per square foot of lead sheet is not exactly its pound designation. In fact, 4 pound sheet lead really weighs 3.65625 pounds per square foot.

The table on the following page shows the relationship between sheet thickness and pound designation as used in the lead trade. Using the table with a few simple calculations will give a rough idea of the amp-hour capacity of your Plante battery design. To do this, you have to know the size of your plates.

For example, for lead plates that are 6"x 7"x $\frac{1}{16}$", multiply 6" times 7" to arrive at the active area of the plate, in this case 42 sq in. The lug surface area and the edges of the plate are not included in these calculations. Then, multiply the active area, 42 sq in, by the plate thickness, $\frac{1}{16}$" (.0625), to get 2.625 cu in, the volume of the lead. Multiply the volume in cubic inches times 6.5 oz (the weight of a cubic inch of lead) to get 17.06 oz. Then, to get the weight in pounds, divide 17.06 oz by 16 oz, which is 1.066 lb.

Refer to the table on the next page to find that $\frac{1}{16}$" thick sheet has an amp-hour capacity of 4.728. Multiply 1.066 lb times 4.728, which is 5.04 amp-hours for 1.066 pounds of $\frac{1}{16}$" thick sheet lead.

When calculating the amp-hour capacity of a battery, use only the positive plates in one cell in your computations. In this particular cell we will have three positive plates, so multiply 3 times 5.04, which is 15.12 amp-hours for the cell. Since the cells will be connected in series, this will be the final amp-hour output of the battery.

Estimating Cell and Battery Output

Solid plate output table

Actual thickness	Industry designation	Actual weight (lbs. per sq. foot)	Approximate thickness	Square inches of surface per pound	Ampere-hours per pound
1/32	2#	1.828	—	157.549	9.457
3/64	3#	2.742	—	105.033	6.304
1/16	4#	3.656	—	78.775	4.728
5/64	5#	4.570	—	63.020	3.783
3/32	6#	5.484	—	52.516	3.152
1/8	8#	7.312	—	39.387	2.364
5/32	10#	9.140	—	31.510	1.891
3/16	12#	10.969	—	26.256	1.576
7/32	14#	12.796	—	22.507	1.351
1/4	16#	14.625	—	19.692	1.182

To calculate the square inches of surface per a given weight of lead, in this case 1.066 lb, simply refer to the table for 1/16" lead sheet to get 78.775 sq in per lb. Multiply 78.775 by 1.11 lb to get 83.97 sq in per plate, or a total area of 251.91 sq in for the three positive plates in the cell.

Design parameters and characteristics of pasted plates

For pasted (Fauré) plates, where active material is applied to a grid support, the amp-hour output is directly related to the amount of surface area of the active material in the plate. The proportions of grid and support to active material can vary. The grid that holds the paste can be very thin as long as it will maintain its strength and rigidity.

Supports made of pure lead need to be more substantial as lead is soft. This is why antimony is alloyed to lead to get stiff and strong support

Pasted plates output table

| Ratio of active material to support || Ampere-hour output of electrode ||
By weight	By volume	Per lb.	Per cubic inch
1:4	1:2	6.0	2.0
1:2	1:1	10.0	3.0
1:1	2:1	15.0	4.0
1.5:1	3:1	18.0	4.5
2:1	4:1	20.0	4.8
2.5:1	5:1	21.0	5.0
3:1	6:1	22.0	5.3
4:1	8:1	23.5	5.4

with less grid material. With antimony added, there can be more space in the plate for active material, which results in higher amp-hour output per volume. Antimony also has less shrinkage than lead upon cooling after casting. So, grids can be either pure lead or lead alloyed with anywhere from 2% to 6% antimony depending on your needs.

To generally calculate the output of pasted plates and the amount of active material required, you have to know the length, width and thickness of the plate, and the volume of the grid spaces. These dimensions will either be per your design or the measurements of a recycled grid.

Calculate amp-hour capacity by volume

As an example, for a 6.75" x 6.3125" x 0.0625" recycled grid, multiply 6.75 x 6.3125 = 42.61 and then 42.61 x 0.0625 to get 2.66 cu in for the plate volume.

There are 261 grid spaces on this plate. Each one is 0.625" x 0.1875" x 0.0625", so to calculate the total grid volume:

    0.625 x 0.1875 = 0.1171875

    0.1171875 x 0.0625 = 0.007324218

    0.007324218 x 261 = 1.9116 cu in

To find the amp-hour capacity of the electrode, calculate the ratio of grid to active material. Since the plate will be 2.66 cu in total, subtract 1.9116 cu in from 2.66 cu in = 0.7484 cu in.

Thus we will have 1.9116 cu in of active material to 0.7484 cu in of lead or lead alloy support grid for each plate. For general purposes, round these figures off upwards to 2:1 ratio. Then, refer to the above table for the 2:1 ratio by volume, understanding that the figure is high because it was rounded off upwards. On the table, this shows an amp-hour rating between 10 and 15 amp-hours per pound (3 to 4 amp-hours per cubic inch).

Calculate weight of active material

To find the weight of the active material needed in each plate, we will use an average weight of 3.25 oz for a cubic inch of active material and multiply this times the volume in cubic inches of the cell spaces:

$$3.25 \text{ oz} \times 1.9116 = 6.213 \text{ oz}$$

So, it will take about 6.2 oz of active material to fill in the grid spaces for each plate. Similarly, you can calculate the weight of pure lead needed to form the grid or the weight of a recycled grid (minus the lug extension) by multiplying 0.7484 x 6.2 oz = 4.64 oz.

Calculate amp-hour capacity by weight

Since the active material provides around 1.875 amp-hours per ounce, you can also estimate the amp-hour capacity by multiplying 1.875 amp-hours x 6.2 oz = 11.625 amp-hours per plate.

Electrolyte

Electrolyte for lead acid batteries consists of sulfuric acid and distilled water. The ratio of sulfuric acid to distilled water determines the specific gravity, or density, of an electrolyte. Specific gravity (sp gr) is measured with a hydrometer, which is a tool that you must have to work with electrolyte.

Measuring specific gravity

Hydrometers are calibrated to the density of water at 60°F, which is sp gr 1.000. Liquid denser than water will show a specific gravity higher than 1.000; and liquid less dense than water will show a specific gravity lower than 1.000.

Since they are calibrated for specific gravity at 60°F, any variation from 60°F at the time of the reading has to be factored in. Thermohydrometers (temperature correcting hydrometers) have adjustment tables built in, or the temperature adjustment can be calculated with instructions that come with a regular hydrometer (shown at left).

Desirable range of sulfuric acid concentration

For most batteries the concentration of sulfuric acid ranges from 28% (sp gr 1.2023), to 36% sulfuric acid (sp gr 1.2684). At about 1.260 the resistance of the sulfuric acid and distilled water electrolyte is at a minimum. Any concentration of sulfuric acid above or

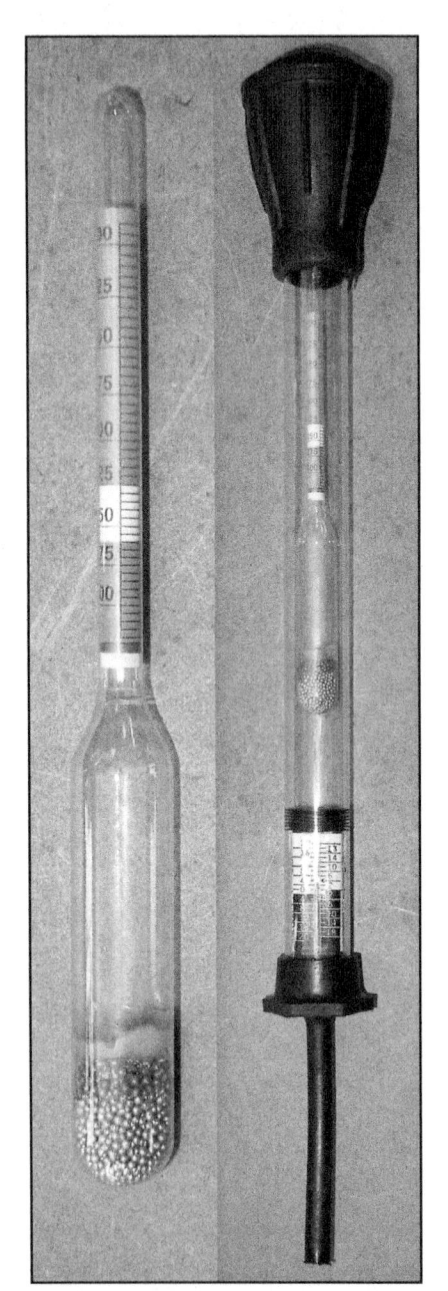

Hydrometer

Electrolyte

below 36% (sp gr 1.260) will decrease the conductance of the electrolyte fluid. So, for best performance in the average battery, the battery acid is usually supplied at 1.265 sp gr. This, as stated earlier, is not the ideal for every situation but works well for a variety of applications.

In a fully charged battery with a 36% concentration of sulfuric acid, the specific gravity will start out at 1.265. During discharge, it is allowed to drop to about 1.150, at which point the electrolyte will contain about 17% sulfuric acid.

When the battery is recharged the sulfuric acid content of the electrolyte will rise again to 36% or 1.265 sp gr. As a battery ages, the electrolyte will have a lower specific gravity when fully charged as the plates deteriorate. A fairly healthy, fully-charged battery will show a voltage of around 12.9 volts after several hours at rest from last charge.

Once you have determined the specific gravity of the electrolyte for your battery, you can outline a rough guide to the relationship between voltage, specific gravity, and percent of discharge.

The basic parameters of battery operation fall between the extremes of full charge and discharge allowance. For all practical purposes I consider a battery fully discharged at about sp gr 1.150. If we start with an electrolyte at sp gr 1.265 (which represents a full charge) and subtract 1.150, we get a spread of 0.115. Divide this by 10 to get the 10% increment for the level of discharge based on the specific gravity of the electrolyte.

To roughly correlate the specific gravity to cell voltage use the following formula:

Specific gravity = open circuit voltage minus (-) 0.845

Open circuit cell voltage = specific gravity plus (+) 0.845

So, if the open cell voltage is 2.1 volts, subtract 0.854, which tells you that the electrolyte specific gravity would be 1.246; or if you have electrolyte sp gr 1.246, add 0.854 which tells you that the open circuit voltage would be 2.1 volts.

37

Electrolyte preparation

Once you have determined the specific gravity and amount of electrolyte that you will need, you will then have to dilute the concentration of acid that you have on hand to arrive at the desired specific gravity in the weight and volume desired.

There are many concentrations available, but the best is probably to get prepared battery electrolyte (sp gr 1.265) from an automotive supplier. This can be used directly without any mixing, which saves the extra time and effort to mix to the desired proportions. It is also safer in that you will spend less time working with electrolyte. Or, if your target is a lower specific gravity as a starting point for your cells, you can cut the prepared electrolyte to your specifications. So, if you can't find prepared electrolyte at exactly your target specific gravity, second choice would to buy it at a slightly higher concentration than your target.

Sulfuric acid can be purchased at any concentration. Higher concentrations can be less expensive, but mixing and diluting can be time consuming. Generally, work with an acid from 93.5% to about 36%, but if you can, work with battery acid available from most automotive parts outlets.

If you purchase acid from an industrial source or chemical supply house, note that there is a range of grades of purity. From highest to lowest purity and cost is ACS, Reagent, USP, NF, Lab, Pure (purified, practical), Technical. Anything from the NF through Technical is fine for battery building, although as with anything else, the purer the acid the better it is for the battery. However, cost is usually a consideration. I find that lower cost Technical grade to be fine in most cases.

Whatever concentration you start with, the graphic at right can be used to determine the proportions needed to reach the desired result.

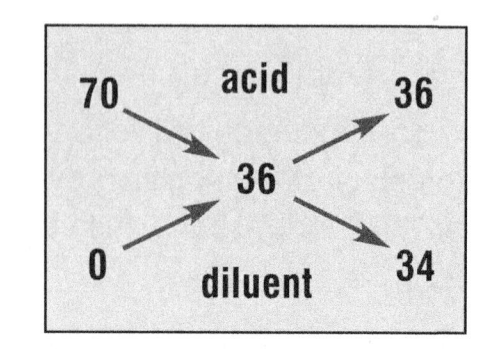

```
70     acid      36
   ↘         ↗
         36
   ↗         ↘
 0    diluent     34
```

Electrolyte

Example: I have a 70% wt solution of sulfuric acid and I need 6 lb of 36% wt acid.

- The target figure (36%) goes in the center of the diagonal lines.
- Place the original solution strength figure (70%) at the top left.
- Place a zero on the lower left diagonal line as we are working with one original solution and are not blending two solutions.
- Subtract the smaller number on each diagonal line from the larger number and place the results on their respective lines. Per our example, subtract 0 from 36 to get 36 and put this figure at the top right. Then, subtract 36 from 70 and place this result (34) in the lower right corner.

The upper right figure is the number of pounds of acid needed. The lower right figure is the number of pounds of distilled water needed. The added total of these two figures equals the total number of pounds (70) that you would get if you add 36 lb of 70% acid to 34 lb of distilled water.

Then, to get the correct proportions to mix to get 6 lb of electrolyte at 36% strength wt:

- Divide the desired weight, 6 lb by 70 lb (the total number of pounds calculated), which equals .0857
- To calculate the amount of acid needed, multiply .0857 x 36, which equals 3.0852 lb of 70% acid.
- To calculate the amount of distilled water needed, multiply .0857 x 34, which equals 2.9138 lb of distilled water.

To convert to parts by volume, divide the weight in pounds by the specific gravity of each liquid. So, per our example:

- 3.0862 pounds divided by 1.6105 equals 1.92 parts by volume of 70% sulfuric acid.
- 2.9138 divided by 1 equals 2.9138 parts by volume distilled water.

Mixing electrolyte solution

When mixing, use an acid resistant rod, such as polypropylene or polyethylene, to slowly stir the electrolyte to aid dispersion. After mixing, test the electrolyte with a hydrometer to determine if any adjustments need to be made.

Wear full safety gear to avoid injury when working with sulfuric acid and electrolyte.

Always add sulfuric acid to the distilled water; or the most concentrated electrolyte to the less concentrated electrolyte. Never add water to sulfuric acid! Always add acid slowly as heat is generated in the mixing process and will spit and spray if combined too quickly. You can mix the electrolyte with an acid resistant rod slowly to aid in the mixing.

Make sure that all mixing and measuring vessels are acid resistant and heat resistant (heat is generated when mixing). Never mix acid and distilled water in metal vessels or use metal utensils to stir electrolyte.

Provide appropriate ventilation in the mixing area and be sure that only authorized personnel have access to the mixing area — no children or pets.

Electrolyte

Determining electrolyte volume

For every amp-hour generated by a lead acid cell, .008 lb of sulfuric acid is broken down and .0066 lb of SO_3 (sulphur trioxide) is removed. To avoid weakening the acid to the point where recharging becomes impossible or extremely difficult, each cell needs to have a sufficient volume of electrolyte.

The volume of electrolyte needed is calculated per active surface area of the plates so that during normal discharge, and by discharge end, the electrolyte's specific gravity maintains above 1.100. The following formula will determine roughly the volume of electrolyte of a given specific gravity, for the active surface area of your plates:

.0066 lb times the output in amp-hours

Calculate the specific gravity range

The output in amp-hours is derived from the calculations for the plates. For instance, for a cell with 3 positive plates that each have an output of 5.04 amp-hours (as in the example given for Planté plates on pages 32-33), the total output is 3 x 5.04 = 15.12 amp-hours. So, 0.0066 lb x 15.12 amp-hours = 0.099792 lb, which is the amount of SO_3 removed.

We will use an electrolyte of 1.265 sp gr, and the electrolyte in any cell should never be reduced to less than 1.100 sp gr no matter what specific gravity we start with. So, the specific gravity parameters for the cell will be 1.265 to 1.100 sp gr.

Calculate the total weight of electrolyte per cell

From the **Specific Gravity Table A** we find the percent of SO_3 nearest to 1.265 and 1.100 sp gr. The nearest specific gravity to 1.100 is 1.1020, which is 12.24% SO_3; and the nearest specific gravity to 1.265 is 1.2684, which is 29.39% of SO_3.

- Subtract 12.24% from 100% = 87.76%
- Subtract 12.24% from 29.39% = 17.15%.
- Divide 87.76% by 17.15 = 5.12
- Multiply 5.12 x 0.099792 lb (the SO_3 removed, from page 41) = .511 lb of electrolyte, the **total weight of electrolyte per cell**.

Specific Gravity Table A

Percent is by weight. Table figures are based on the density of water at 4°C as unity.

Percent H_2SO_4	Baume	Specific gravity	Percent SO_3
1	0.7	1.0051	0.8163
2	1.7	1.0118	1.633
3	2.6	1.0184	2.449
4	3.5	1.0250	3.265
5	4.5	1.0317	4.082
6	5.4	1.0385	4.898
7	6.3	1.0453	5.714
8	7.2	1.0522	6.531
9	8.1	1.0591	7.347
10	9.0	1.0661	8.163
11	9.9	1.0731	8.979
12	10.8	1.0802	9.796
13	11.7	1.0874	10.612
14	12.5	1.0947	11.43
15	13.4	1.1020	12.24
16	14.3	1.1094	13.06
17	15.2	1.1168	13.88

Percent H_2SO_4	Baume	Specific gravity	Percent SO_3
18	16.0	1.1243	14.69
19	16.9	1.1318	15.51
20	17.7	1.1394	16.33
21	18.6	1.1471	17.14
22	19.4	1.1548	17.96
23	20.3	1.1626	18.78
24	21.1	1.1704	19.59
25	21.9	1.1783	20.41
26	22.8	1.1862	21.22
27	23.6	1.1942	22.04
28	24.4	1.2023	22.86
29	25.2	1.2104	23.67
30	26.0	1.2185	24.49
31	26.8	1.2267	25.31
32	27.6	1.2349	26.12
33	28.4	1.2432	26.94
34	29.1	1.2515	27.75

Electrolyte

Percent H_2SO_4	Baume	Specific gravity	Percent SO_3
35	29.9	1.2599	28.57
36	30.7	1.2684	29.39
37	31.4	1.2769	30.20
38	32.2	1.2855	31.02
39	33.0	1.2941	31.84
40	33.7	1.3028	32.65
41	34.5	1.3116	33.47
42	35.2	1.3205	34.29
43	35.9	1.3294	35.10
44	36.7	1.3384	35.92
45	37.4	1.3476	36.73
46	38.1	1.3569	37.55
47	38.9	1.3663	38.37
48	39.6	1.3758	39.18
49	40.3	1.3854	40.00
50	41.1	1.3951	40.82
51	41.8	1.4049	41.63
52	42.5	1.4148	42.45
53	43.2	1.4248	43.26
54	44.0	1.4350	44.08
55	44.7	1.4453	44.90
56	45.4	1.4557	45.71
57	46.1	1.4662	46.53
58	46.8	1.4768	47.35
59	47.5	1.4875	48.16
60	48.2	1.4983	48.98
61	48.9	1.5091	49.80
62	49.6	1.5200	50.61
63	50.3	1.5310	51.43
64	51.0	1.5421	52.24
65	51.7	1.5533	53.06
66	52.3	1.5646	53.88
67	53.0	1.5760	54.69

Percent H_2SO_4	Baume	Specific gravity	Percent SO_3
68	53.7	1.5874	55.51
69	54.3	1.5989	56.33
70	55.0	1.6105	57.14
71	55.6	1.6221	57.96
72	56.3	1.6338	58.77
73	56.9	1.6456	59.59
74	57.5	1.6574	60.41
75	58.1	1.6692	61.22
76	58.7	1.6810	62.04
77	59.3	1.6927	62.86
78	59.9	1.7043	63.67
79	60.5	1.7158	64.49
80	61.1	1.7272	65.31
81	61.6	1.7383	66.12
82	62.1	1.7491	66.94
83	62.6	1.7594	67.75
84	63.0	1.7693	68.57
85	63.5	1.7786	69.39
86	63.9	1.7872	70.20
87	64.2	1.7951	71.02
88	64.5	1.8022	71.84
89	64.8	1.8087	72.65
90	65.1	1.8144	73.47
91	65.3	1.8195	74.28
92	65.5	1.8240	75.10
93	65.7	1.8279	75.92
94	65.8	1.8312	76.73
95	65.9	1.8337	77.55
96	66.0	1.8355	78.37
97	66.0	1.8364	79.18
98	66.0	1.8361	80.00
99	65.9	1.8342	80.82
100	65.8	1.8305	81.63

Calculate the volume of electrolyte per cell

Refer to **Specific Gravity Table B** and choose the closest figure to the concentration of acid you are working with either by weight or specific gravity. There are 1,728 cubic inches in a cubic foot, so divide your chosen weight by 1,728. The result will be the weight of a cubic inch of electrolyte at the specific gravity you have chosen to use in your cells.

For example, to use a 1.265 sp gr electrolyte, choose the closest figure in **Table B** to the specific gravity desired, in this case 1.2609 sp gr with a weight of 78.64 lb per cubic foot. Divide 78.64 lb by 1,728 = .0455 lb per cubic inch. Then, divide .511 lb by .0455 = 11.23 cu in, the **volume of electrolyte needed for each cell**.

Specific Gravity Table B

Percent is by weight. Specific gravity figures were determined at 60°F, compared with water at 60°F.

Specific gravity	Baume	Percent	Weight of 1 cu. ft. in lbs. avoirdupois
1.0000	0	0	62.37
1.0069	1	1.02	62.80
1.0140	2	2.08	63.24
1.0211	3	3.13	63.69
1.0284	4	4.21	64.14
1.0357	5	5.28	64.60
1.0432	6	6.37	65.06
1.0507	7	7.45	65.53
1.0584	8	8.55	66.01
1.0662	9	9.66	66.50
1.0741	10	10.77	66.69

Specific gravity	Baume	Percent	Weight of 1 cu. ft. in lbs. avoirdupois
1.0821	11	11.89	67.49
1.0902	12	13.01	68.00
1.0985	13	14.13	68.51
1.1069	14	15.25	69.04
1.1154	15	16.38	69.57
1.1240	16	17.53	70.10
1.1328	17	18.71	70.65
1.1417	18	19.89	71.21
1.1508	19	21.07	71.78
1.1600	20	22.25	72.35
1.1694	21	23.43	72.94

Electrolyte

Specific gravity	Baume	Percent	Weight of 1 cu. ft. in lbs. avoirdupois
1.1789	22	24.61	73.53
1.1885	23	25.81	74.13
1.1983	24	27.03	74.74
1.2083	25	28.28	75.36
1.2185	26	29.53	76.00
1.2288	27	30.79	76.64
1.2393	28	32.05	77.30
1.2500	29	33.33	77.96
1.2609	30	34.63	78.64
1.2719	31	35.93	79.33
1.2832	32	37.26	80.03
1.2946	33	38.58	80.74
1.3063	34	39.92	81.47
1.3182	35	41.27	82.22
1.3303	36	42.63	82.97
1.3426	37	43.99	83.74
1.3551	38	45.35	84.52
1.3679	39	46.72	85.32
1.3810	40	48.10	86.13
1.3942	41	49.47	86.96
1.4078	42	50.87	87.80
1.4216	43	52.26	88.67
1.4356	44	53.66	89.54
1.4500	45	55.07	90.44
1.4646	46	56.48	91.35
1.4796	47	57.90	92.28

Specific gravity	Baume	Percent	Weight of 1 cu. ft. in lbs. avoirdupois
1.4948	48	59.32	93.23
1.5104	49	60.75	94.20
1.5263	50	62.18	95.20
1.5426	51	63.66	96.21
1.5591	52	65.13	97.24
1.5761	53	66.63	98.30
1.5934	54	68.13	99.38
1.6111	55	69.65	100.48
1.6292	56	71.17	101.61
1.6477	57	72.75	102.77
1.6667	58	74.36	103.95
1.6860	59	75.99	105.16
1.7059	60	77.67	106.40
1.7262	61	79.43	107.66
1.7470	62	81.30	108.96
1.7683	63	83.34	110.29
1.7901	64	85.66	111.65
1.7957	64¼	86.33	112.00
1.8012	64½	87.04	112.34
1.8068	64¾	87.81	112.69
1.8125	65	88.65	113.05
1.8182	65¼	89.55	113.40
1.8239	65½	90.60	113.76
1.8297	65¾	91.80	114.12
1.8345	66	93.19	114.47

REBUILDING and RECYCLING BATTERIES

Reconditioning and Rebuilding Batteries

Reconditioning vs. rebuilding

Reconditioning discarded batteries can consist of slow charging, total electrolyte replacement, water cure, and/or bringing electrolyte up to (or down to) the required specific gravity. It can also include pulse charging and the addition of chemical compounds.

Rebuilding can involve repairing cracked cases, replacing missing or broken caps, burning in broken terminals, and/or any operation that requires removing the case cover to perform operations on internal components.

My estimate is that 25% of discarded batteries are fine and simply require a slow charge to get them back in shape. This means that at least two out of every ten discarded batteries will be totally serviceable for at least another year or two. There are many reasons for this. Many people mistakenly replace a battery thinking that there is something wrong with it when, in fact, their charging system is malfunctioning or the connections are corroded and or loose. Some automotive repair shops will recommend battery replacement after a certain amount of time even though the battery may be performing quite well. So, with the investment of little time and effort, you can find perfectly good batteries that just need to be charged. Another 10% of batteries are capable of being rejuvenated by water treatment, addition of new electrolyte and slow charging. The remaining 65% need to be rebuilt to be functional, or they can be salvaged for parts.

Assessing discarded batteries

If you are discriminating at the point of pickup you will wind up with a higher percentage of usable batteries for your efforts. Here's a checklist for assessing them:

Reconditioning and Rebuilding Batteries

1. Look for liquid or a wet spot where the battery is sitting. This is an indication of a leaky case.
2. Are the battery's terminals intact, and the caps all there?
3. If the case is bulged, avoid that battery.
4. Open the battery caps and see if there is electrolyte in the battery cells. If there is no electrolyte in the cells, or if the electrolyte does not entirely cover the plates, the plates are probably pretty far gone. It may be possible to recover them, but it would probably take a lot of time and effort. Unless there are other parts worth salvaging, I would avoid such a battery.
5. Check the open circuit voltage at the terminals with a multi-meter. If you have a large number of batteries to choose from, only take those that read 11 volts or more. This can be a fooler as there are a lot of battery problems that this test does not address, but the multi-meter test boosts the chances that the batteries you choose will only need to be charged to be put back in service. Of course, if you have fewer batteries to choose from you can make do and apply your skills to recondition the battery as best you can.

Reconditioning

Unfortunately, some problems are not visible, such as internal problems, or external problems such as micro cracks in the case. Micro cracks are very thin invisible cracks in the case that you can't detect until you charge the battery. At that point the battery seems to sweat and it gets wet around the base of the battery. This defect only shows up while charging as there is more pressure within a cell from the buildup of gas before it is released through the vent caps. Batteries that have been severely weathered by sitting outside for too long are apt to have micro cracks, and most of the cases that have this defect are also bulged, so it is good practice to avoid bulged cases if you can.

If you do have micro cracks, note where they are, and clean the outside of the case. Then, rub the leaky areas of the case with Viton® caulk and apply a thin coating of epoxy over the Viton® for a durable seal. This will usually solve the problem if you do a thorough job.

When you get a battery back to your shop, check the electrolyte level. Make sure the plates are totally covered by electrolyte. If they are not, add distilled water to about ¼" above the tops of the plates.

Battery reconditioning by charging

Charge the battery on a slow charge at low rate for 24 hours. After the first few hours, take a specific gravity reading and record that. Do not use a fast charger. This will just overheat the battery and you will get nowhere fast. It's much better to take your time. During the charge cycle, notice if the battery overheats. It should not under any circumstances be allowed to go above 110°F. If the battery gets hot, take it off the charger immediately, let it cool down and try charging again. If the battery overheats again, it should be either rebuilt or discarded.

Several times during the first 24 hour charge, note the specific gravity at intervals and record the readings. At the end of the 24 hour period, turn off the charger and let the battery sit for an hour.

Make a final specific gravity reading. Is the specific gravity higher than the first reading, and has the voltage increased?

An hour after the first charge, if the battery has a reading of 12.9 to 13.2 volts, it can be put into service as is, or put on a maintenance charge to make it ready for use. If the battery voltage is between 12 and 12.8 volts, the battery can be transferred to a smart (automatic charger) to continue charge for 24 hours.

If you are charging a second time to bring up the 12 to 12.8 volt range battery, let the battery rest an hour after the second charge, then measure the voltage and specific gravity. If the battery voltage

Reconditioning and Rebuilding Batteries

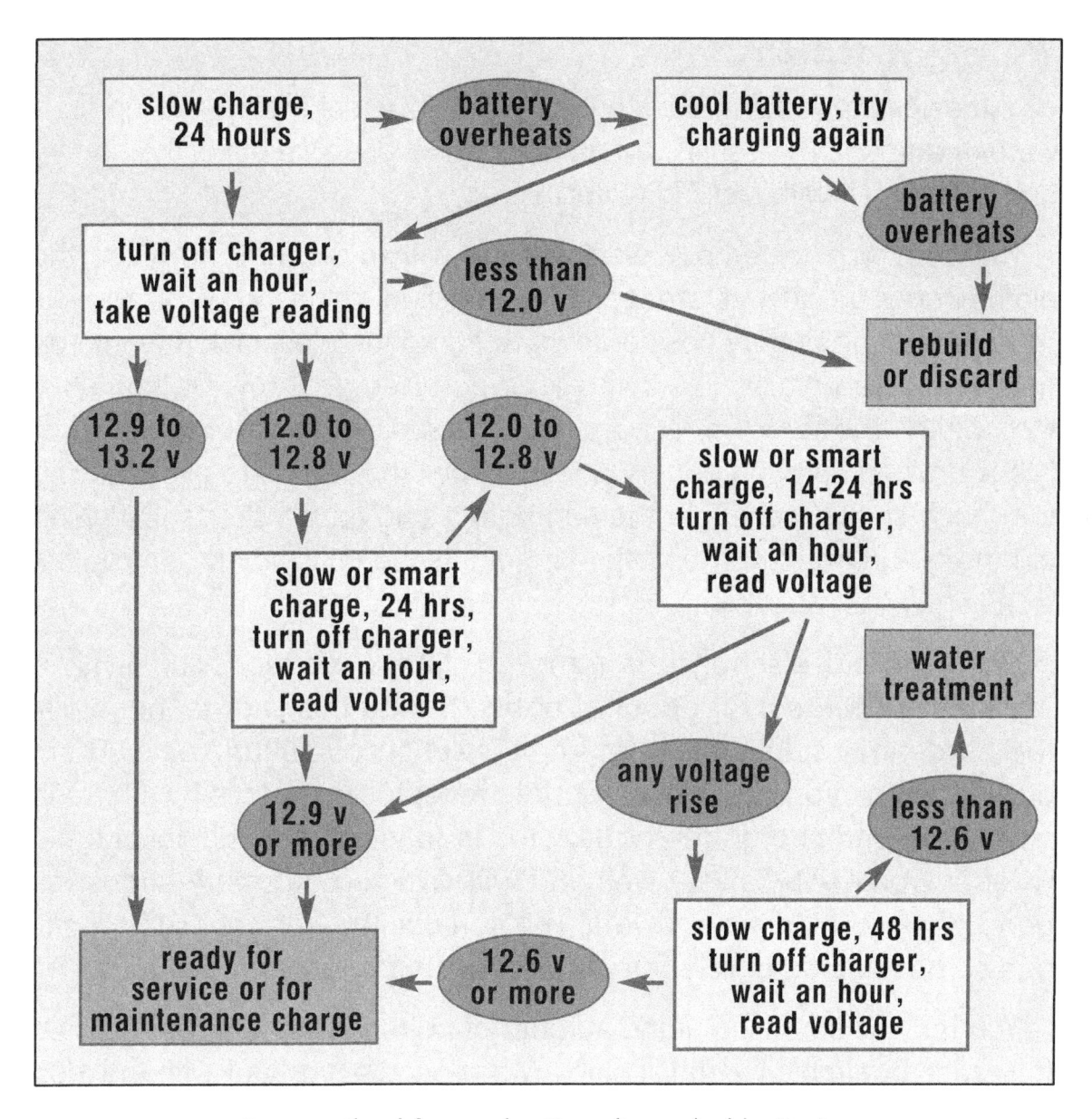

One method for evaluating discarded batteries

has risen to at least 12.9, the battery is OK for service and maintenance charge. If the battery is below this voltage, charge it for another 14 to 24 hours to see if it will rise any further. If you get a rise, even if it's minuscule, you can try up to 48 hours more of slow charging to see what it will do. After this amount of time (about 5 days of charging) you can put the battery aside that is not progressing for the next phase.

Water treatment

If the voltage reading does not get up to 12.6, set the battery up for a water treatment. This consists of dumping the electrolyte out of the battery and replacing it with distilled water.

The best way to remove electrolyte is with a syphon, however, the plates in many batteries are too closely packed for this to work. You will have to turn the battery upside down to dump the electrolyte out into an appropriate receptacle, like a kitty litter pan (see page 61). Do not leave the battery upside down for very long. As soon as most of the electrolyte is drained, turn the battery right side up. The idea is to avoid having too much sludge penetrate the separators and cause shorts. Dumping out the electrolyte can cause shorts, but it is the only way to remove it if siphoning is not possible.

Once the old electrolyte is removed, fill the battery with distilled water, and charge it for about 14 hours. During this charge, hopefully the hardened sulfates will return to solution as sulfuric acid. Then, dump this solution into an appropriate receptacle, replace it with fresh electrolyte and charge the battery for 14 to 24 hours. This technique, when it works, works very well and will give you a battery almost as good as new. If it does not work, you can use the battery as is if it still has some life in it, or relegate it to rebuilding or salvage for parts.

You may be surprised at the number of batteries that can be returned to service by simple reconditioning. Many people start out with the idea of rebuilding batteries, but when they find out how many batteries can be brought back to service without even cracking open a case they choose not to go any further.

Rebuilding batteries

If a battery rebuild is necessary, consider what lengths you are willing to go to rebuild a battery. If you are simply repairing a cracked case or cover, or replacing a cap, or burning on a new terminal, this requires only minimal time and effort. If the case needs to be opened to discern the problem or problems and then repair, you will need the full range of skills required for battery building.

Battery malfunctions

Battery problems can be simple or complex. During service, batteries lose active matter from their plates, the plates corrode and crack, the separators are bridged by conductive trees and the plates short out, plates warp, buckle and bulge. Other problems can occur such as intercell connectors breaking and plates coming loose from straps.

To discern what the problem is, drain the battery (see page 61) and take the cover off (see page 64).

Once the cover is removed, first look for any readily visible problems. The intercell connections should be solid. Test for continuity between the intercell connects. If that checks out well, look for any visible breaks between the straps and plates along the points of connection or further down. If you can see no problems, cut the intercell connects, pull the elements from the cells and pull the positive plates from the negative group.

After the elements and groups are pulled, remove the separators from the plates to view the surface of the plates. The plate surface should be intact without any loss of active material or parts of the grid missing.

If the plates are in good shape, probably the plates have bridged, causing a short through the separators. Replace the separators with new ones, reinsert the plates back into the cells, and reburn or screw fasten the intercell connection.

If the positives or negatives break away from the straps while you are pulling them apart then you have probably discovered the problem. Usually the positives rot away and the negatives are fine. If this is the case, insert a new set of positives.

In some cases, just one plate needs to be replaced. If this is the case, hacksaw a slot where the plate was burned to the strap, and burn an new plate in on the plate burning rack. If the whole group needs to be replaced, hacksaw slots in the strap for the insertion of new plates, and burn them in on a burning rack.

Replace plates with new or renovated plates that are the same type and same size, or very close to it. For pasted plates, recycled grids from discarded batteries can be reworked and repasted. All separators should be replaced.

In some cases, sludge has built up above the ridges that hold the plates in the cell well. This causes shorts. However, once you have turned the battery upside down to drain the acid, you will not be able to tell if this was the cause of the battery's failure.

If you have pulled the elements, take this opportunity to remove any sludge on the bottom of the case, and give the whole inside of the case a thorough cleaning before replacing the elements.

There may be bulged, warped, or buckled plates. It may be possible to gently work them back into shape. If they are not too badly deformed, leave them alone. If they are deformed in a major way, pull the groups apart and remove the separators. At this point, check to see if much of the active material has fallen from the plate into the well due to the buckling. If this is so, cut the plate or plates from the strap and burn new plates to the strap.

If you find the plates intact after the separators are removed, place plywood inserts between the plates and on the outside ends of the group. When these are in position, take a carpenter's clamp and slowly compress the plates back into shape. This may or may not work. Bending the plates back into shape may loosen the active material in the grids and it may crumble out.

Another option is to straighten the plates on a small hydraulic press. This would tend to add more pressure to compress the active material back into the grids. Care must be taken not to distort the plates with excessive pressure. If this operation is successful, replace the separators and reinsert into cells wells.

Reconditioning and Rebuilding Batteries

To rebuild batteries, you will at the minimum need to disconnect and reconnect intercell connectors, replace separators and clean the cases from sludge.

If you replace plates and whole plate groups, you will need to slot the strap and burn the plates to the strap on the plate rack, and you will have to fabricate the replacement pasted grid.

While doing a rebuild, keep any plates, groups and elements that you pulled and are not working on at the moment in a bucket of water so that they do not dry out. All the skills and techniques needed for rebuilding batteries, such as burning a plate to a strap and pasting plates, are covered in detail elsewhere in this book.

Recycling Battery Parts

Most of the materials needed for building lead acid batteries can be obtained from discarded lead acid batteries. From discarded batteries you can recycle:

- Cases, case lids and cell well caps
- Lead posts and straps, intercell connectors and straps
- Plate grids
- Sulfuric acid

All batteries and battery banks wear out and have to be replaced at some time, and people generally don't want the old ones hanging around. If you offer to remove them, most people will be grateful, as this will save them the effort of taking them to the local recycling station and paying a fee for proper disposal.

A simple local classified ad will probably get you what you need. If you do put out an ad, specify the type of battery that you are willing to remove. The type of battery that you want will, of course, depend on what you plan to build.

Sealed lead acid batteries do not have the type of covers and design features that you probably need for building flooded lead acid batteries. They could be useful for certain projects, but I suggest staying with flooded lead acid batteries at first. There are many sources of spent batteries. If you are looking for deep cycle batteries, talk with:

- People with solar or wind electrical systems
- Solar and wind system installers
- Golf course maintenance managers for golf cart batteries
- Factory maintenance managers for fork lift and other types of industrial batteries

Recycling Battery Parts

For automotive batteries, contact:

- Automotive and metal scrap yards
- Gas stations
- Automotive repair shops
- Recycling stations
- Individual automobile or truck owners

You can also inquire about mid-range marine type batteries at your local marine service shop if you happen to live near the water.

Usually deep cycle or industrial batteries have larger sized cell wells than automotive batteries. Many deep cycle batteries come in three cell 6 volt versions, and most automotive batteries come in six cell 12 volt versions. If it makes a difference to you, focus on the businesses or people that are likely to have exactly what you need.

Actually you can reuse any type of case, as long as there is enough room for the plates you plan to use. If the case is too big, you can use shim material to fill the extra space. Lids and caps can also be reused.

You can learn a lot by obtaining different types and brands of batteries to see how they are constructed, and how easy or difficult they are to work with. Some case lids are easy to remove, while others are more difficult.

What battery parts can be reused or recycled?

From each battery, you can recycle the lead from the straps, intercell connectors and posts. They may be either reused as is, or melted down and molded into new pieces. Construction of plates varies a lot, so you will have to see what you find when you open a battery for assessment.

Although it's theoretically possible to recycle the lead from pasted or solid negative plates (the negatives are usually much more intact than the positives), it is not advisable because of the chemical layers on the plates. These chemicals would have to be burned off, which would produce toxic air pollution that can adversely affect your health and that of others. The only safe way of doing this is to use a fume hood with ventilation and filters to remove the particulates.

However, the grids from negative pasted plates can be recycled by poking out the old pasted oxides inside the grids, and repasting the grids. These plates can be used for both positive and negative pasted plates in a new battery.

Many positive plates are tubular, and are not amenable to recycling because they are usually rotted.

Separators cannot be reused as they contain too many contaminants, are uneven, and when the batteries are turned upside down to drain, the sludge from the bottom of the wells will get lodged in the separator material. Separators should be made from new material when you build or rebuild a battery.

Battery electrolyte can be recycled. However, electrolyte specific gravity will vary from battery to battery, and you will have to adjust the specific gravity after initial processing to suit your needs. Using recycled electrolyte works well enough, but fresh electrolyte is better as you will be assured of having no contamination. Impurities in the electrolyte can reduce a battery's performance.

Recycling lead for battery building

For battery plates or any other lead parts, recycled sheet lead can be used. Lead sheet is used for sound insulation in construction as well as for roof flashing. Most large metal salvage yards will have lead sheet that you can purchase. Demolition and construction contractors may be another source. Salvaged sheet lead is usually unalloyed but can contain other metals such as antimony in small amounts. This may or may not suit your needs. With recycled materials you will nearly always be working with a unknown. Most salvage workers will not know whether the lead is alloyed or not alloyed.

Tradespeople who know lead designate unalloyed lead as soft lead, and alloyed lead as hard lead. Most sheet lead is designated soft lead.

Recycling Battery Parts

For its most common applications, sheet lead needs to conform to surfaces, deaden sound and be impermeable to x-rays, thus it does not need a hardening element, and is usually unalloyed.

On the other hand, automotive battery components are mostly hard lead and may be alloyed with a variety of metals with a large percentage of antimony as the alloy. Antimony adds stiffness and is more brittle than unalloyed lead. It gives strength to fragile thin grids and also makes it easier to cast small or thin parts. Alloys of lead and antimony will not shrink as much as pure lead when cooled after casting. The alloys are also more resistant to deterioration from the action of the electrolyte in the battery cell.

Other types of batteries (non-automotive) vary between unalloyed lead and alloyed lead depending on the purpose of their design. The degree of antimony alloy in any lead product will usually not be more than 12%. Most automotive batteries have an antimony alloy content of from 2% to 6%.

Used wheel weights are another abundant source of lead for casting battery components. They are available at most salvage yards, automotive centers that change tires for customers, and garages in general. Wheel weights can have an alloy content of from .75% to 2%, but there's really no way to know for sure. When it comes to assessing salvaged lead the only person who will really know the composition of the material at hand is the original manufacturer.

For group straps, posts, and intercell connectors, lead antimony alloy is fine for Plante or Fauré plates. If you are casting grids or using recycled grids for Fauré type pasted plates, the lead antimony alloy will work well. However, for Plante type plates, use soft lead for the plates proper. A low antimony alloy may work with Plante plates, but I have never tried it.

You may be surprised at where you will find lead. Above is a small part of an antique water pipe system consisting of lead pipe and hand carved wooden connectors that I dug up in my back yard.

Handling used batteries

When collecting batteries and moving batteries around, use a wheel cart. Batteries are heavy and can hurt you badly when lifted improperly. Learn how to lift them appropriately to avoid injury. Calculate your moves before you make them so that you do as little lifting as possible.

Make sure that you are fully suited when working with batteries and that your workspace is well set up for the task. Recycling outdoors is the best option as that will afford you plenty of ventilation and space.

Parts of your work area should be covered with plastic to hold parts and debris, especially when you extract plates from cases.

Keep a few 5 gallon buckets full of water on hand and have plenty of baking soda to neutralize acid on plate groups. Have one bucket to dip and wash acid from your gloves before you take them off. Several small plastic containers filled with water and baking soda can also hold smaller parts such as straps cut from plates.

Draining the electrolyte

You will need a plastic receptacle for draining the battery acid, along with a couple of bricks or plastic blocks so that when the battery is turned upside down to drain, the battery case will be raised above the bottom of the container and the liquid will drain out away from the case freely. The blocks should be spaced so that the overturned battery will be supported

Recycling Battery Parts

Battery turned upside down to drain the electrolyte from the case

firmly and will not fall off while draining. The plastic container that will receive the drained acid also needs to be super stable and firm so that it will not flip over on you if you accidentally drop a battery on the container's edge.

Before you set up draining apparatus, think through the process thoroughly and plan the moves involved. Consider the what-if's so that you take care of any problem before it happens. You can use a large cat or dog litter box set in sand so that it can not be tipped over. There are many other options that you can devise to make your operation safe.

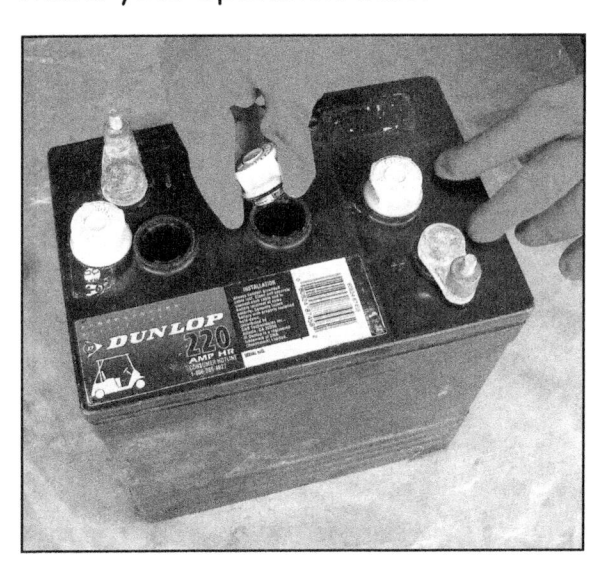

◆ When everything is ready and in place, unscrew the caps from the battery cover and place the caps in a plastic container for cleaning.

61

The Battery Builders Guide

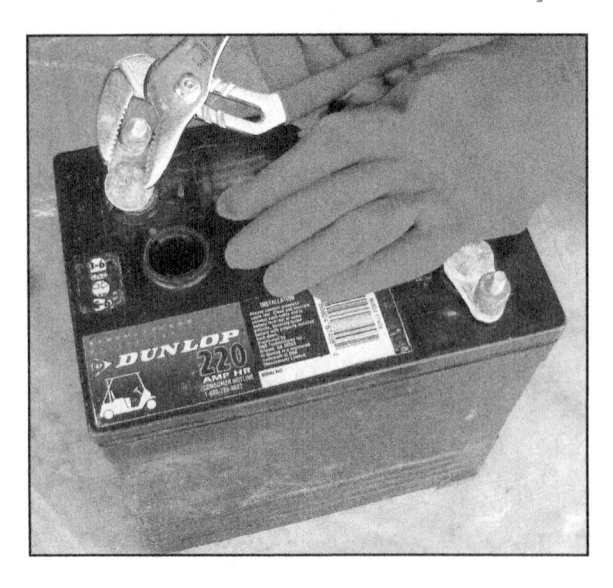

◆ Next, twist off the terminals with a pair pliers. Terminals are easily removed in this manner. Place the terminals in a receptacle for cleaning and melting. Grab the battery by the sides, lift it, and slowly turn it over into the container and onto the bricks so that it will drain. Leave the battery in this position for about a half hour to an hour.

◆ Grab the battery by the sides, remove the battery from the bricks and turn it right side up again onto a plastic sheet.

◆ Wipe any battery acid from the battery case with a rag soaked in a baking soda and water mix.

During these various operations it's a good idea to dip your gloved hands into a bucket of water with baking soda, reserved for removing any traces of acid on your gloves. It's not absolutely necessary but it will help you to avoid contaminating (or re-contaminating) the outside of your case with electrolyte. As you work, keep the components as clean (meaning acid free) as possible in each step.

62

Recycling Battery Parts

- Next, pour the acid from the plastic container into a polypropylene or polyethylene bottle. If you are going to recycle the sulfuric acid, let the bottle sit undisturbed for about 3 days. This will allow any sediment to collect at the bottom of this settling bottle.

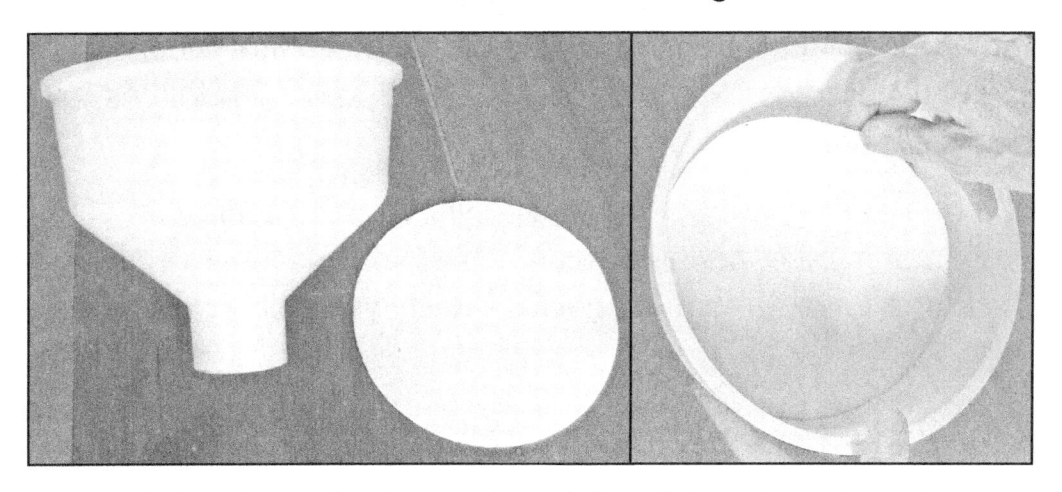

Cut a piece of polypropylene felt to fit in a large funnel for filtering the used electrolyte

- After the residue settles to the bottom, you can pour this acid into a storage bottle through a large size funnel with a filter cut from polypropylene felt.

If you are going to process electrolyte from more batteries, when pouring from the settling bottle, do not pour the residue at the bottom of the bottle onto the filter — leave a small amount of acid with the residue in the settling bottle. You can use the settling bottle to store residue-laden sulfuric acid from your other batteries, or you can just properly dispose of it and not bother to filter further.

When you are finished filtering used electrolyte, you can filter out the residue-laden acid in the settling bottle. Then, take the filter out of service to be disposed of or recycled in an appropriate manner.

Bottles for storing sulfuric acid and their lids need to be sulfuric acid resistant. Always label and store acid in a safe place that is not accessible to unauthorized personnel, and follow MSDS recommendations for storing this substance.

Test the specific gravity of the electrolyte in each bottle that you store and mark that number on the outside of the bottle so that later if you have to adjust for a lower or higher specific gravity you will have a number to start with.

Opening the battery case

Once the electrolyte has been drained out of the battery case, the lid can be removed from the case. There are several ways to do this — you can use heat, such as a hot knife or a soldering iron with a knife tip; or just mechanical force (chisel, putty knife, hammer or whatever). Battery manufacturers use a wide range of adhesives and sealing techniques, so the best method to use will depend on the cover and how it was bonded to the case. Depending on the method you decide to use, figure out your best angle of attack, and either turn the battery on its side or leave it upright.

The mechanical approach

Start with a thin but stiff putty knife that can be wedged into the space between the case and the cover and then tap gently with a hammer to cut into the sealant along the edges. Do this all the way around the battery cover rim and then test for movement. Go around the rim again, tap a little harder and cut deeper into the sealant holding the case and cover together. You will have to do this process a number of times, but after a while the cover will loosen and at certain points you can push up on the cover. Once you find one of these points, start from there and move outward to the rest of the rim until the cover is totally loose around the rim.

Recycling Battery Parts

You may notice that the cover will still not come free from the case. The reason for this is that sealant is also used along the top of the cell walls and around the terminal posts to seal the cover.

You can apply pressure upward on the cover at this point with a chisel, or putty knife or whatever to crack the final bond. This is a delicate operation. Work around the case cover and do a little bit on each side. After a few round trips you should be able to forcefully spring the cover from the case. If you get a crack starting either on the rim of the cover or on the top of the cover, move to a different area and work there for a while. Try not to make any major cracks in the cover. Minor cracks can be mended and parts glued back together again, but it is best to do as little damage as possible.

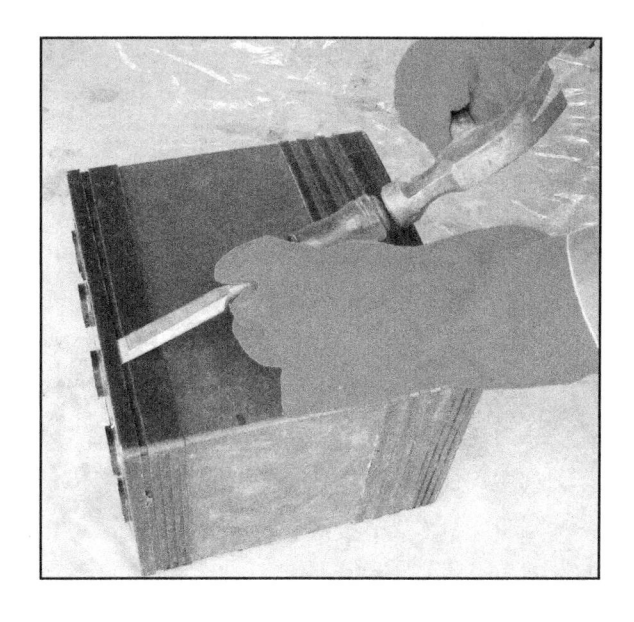

Applying heat

To use a hot knife or iron during all or part of the process, use a low heat of around 375°F at first. Wood burning irons with knife tips can be a lot hotter and will melt the case cover quite fast. If you work with these, you must time yourself well and use swift movements.

When using a heating device, move along the edge and then pry up with another tool such as a putty knife so that the lid doesn't re-bond as the sealant cools. Two people working with this technique can do the best job.

Assess the condition of the lid

After removing the case cover, assess any damage and decide whether it can be repaired adequately. If it will be reused as is or repaired, the lid should be immersed in a bath of water and baking soda to neutralize any acid. Wipe the lid clean of any debris, and remove any hardened lumpy adhesive that still clings to the edges of the cover so that it will be as smooth as possible.

Removing the battery elements

Once the lid is off the case, remove the elements from the battery. If they have intercell connections, the elements will be attached to each other through the cell wells. To get the elements out of the cells you need to cut the intercell connecting rods. I usually use a thin chisel to bend the strap attachments away from the cell walls so that I can get at the connecting rods with the chisel. Then, hammer the chisel gently to sever the connecting rod between the intercell connectors. However you can do it, each element needs to be disconnected from the others so that you can remove it from the cell well.

Once the intercell connectors are cut, pull each element from the case. In some batteries the plates will have expanded so much that they become tightly retained in the case.

An element that has been pulled out of the cell well.

Recycling Battery Parts

To test for removal, grab both of the group straps and pull straight upward. If they do not budge you may need pliers to grab the group straps. If the case tends to move upwards as you pull the plates, you can use your feet to hold it in place while you lift the element.

Lay the elements on a plastic sheet as each is removed.

If you are not going to use any components from the groups, put them in a 5 gallon bucket labeled "battery components."

If you are going to use just the straps, posts, and intercell connects, cut them off with a metal shear and then put the group plates in the "battery component" bucket. Immerse the parts that you will reuse in a solution of water and baking soda to neutralize the acid.

Pull the groups of plates apart

Separating the groups of plates

If you are going to use the negative plates they must be separated from the positive groups. To do this, grab each of the straps in an element and pull the two groups apart.

In most cases the positive plates will break and crumble from the straps, and you will have to clean the positive plate materials out with a chisel, putty knife or other similar tool to remove the positive plate material from in between the negative plates. Cut the straps from the positives with metal shears. Put the straps into a container with water and baking soda to neutralize the acid

Right, removing crumbled positive plate material from in between the negative plates.

Left, cutting the strap from the positive group

Below, separated groups, negative groups in the center, positive groups on the right

Recycling Battery Parts

What is left of the positives should be put into a 5 gallon bucket labeled "battery components" and put aside for the moment, to be recycled later. Put the negative plates into a bucket with a water and baking soda solution to neutralize the acid. The solution will fizz and bubble a lot and then subside after a period of time. Pour more baking soda in, and the water will begin to fizz again. Leave the plates in this solution for a day or two before removing.

At this point there will be several containers of baking soda solution with parts in them being neutralized, and several containers labeled "battery components" with the parts that will be recycled.

The buckets of components that you will not use can be brought to your local recycling station, and they will be treated as any other battery for disposal. You could also have a local recycler pick up your materials for a fee. Whatever you do, make sure to dispose of all recyclables in accord with your local regulations and requirements.

If you are disassembling a battery and are going to reuse the case, get the battery's plate dimensions. Be sure to include measurements of tab lengths, distance of top of strap to plate, etc, so that you'll know the exact dimensions for plates and straps when you replace them.

Soaking a group of negative plates in water and baking soda.

Below, group of negative plates

Cleaning the case

A thin firm stick or thin spatula can be used to remove the residue on the bottom of the cell wells. Follow this up with a rag tied to the end of a stick to soak up and catch any remaining residue. Next, the cases should be cleaned with a solution of water and baking soda to neutralize the acid. Thoroughly remove all traces of residue. The cases need to be very clean for further use.

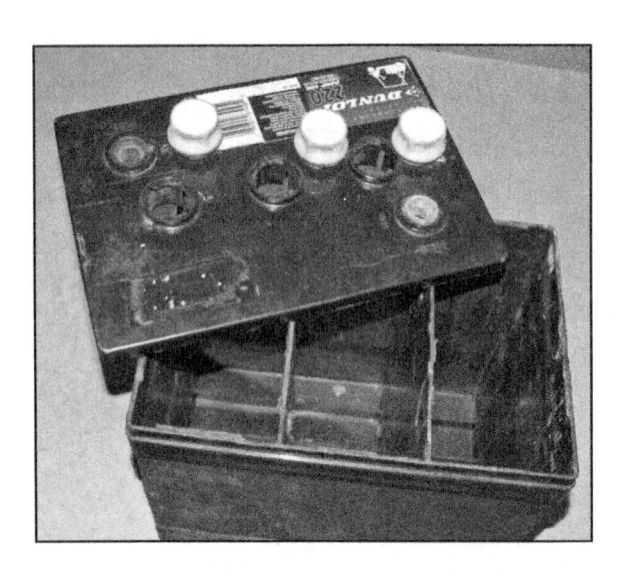

Repairing case covers

Most cracked case covers can be repaired easily with polypropylene glue sticks and hot melt gun. It does not look pretty, but it bonds and seals well. Pieces that have broken off can also be reattached using the glue gun.

Recycling Battery Parts

A hot melt gun can be used for lid repairs

Processing negative plates for reuse

Remove the plates from the neutralizing solution. Even though you have soaked the negatives in a neutralizing solution there will still be acid residue on the plate grid, so it is best to handle plates with appropriate gloves at all times.

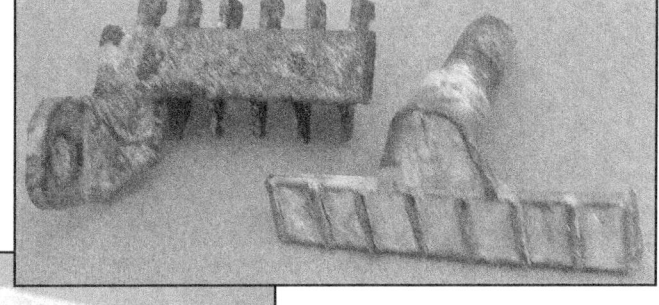

Above, straps that have been cut away from plate groups.

Left, a group of negative plates

71

The Battery Builders Guide

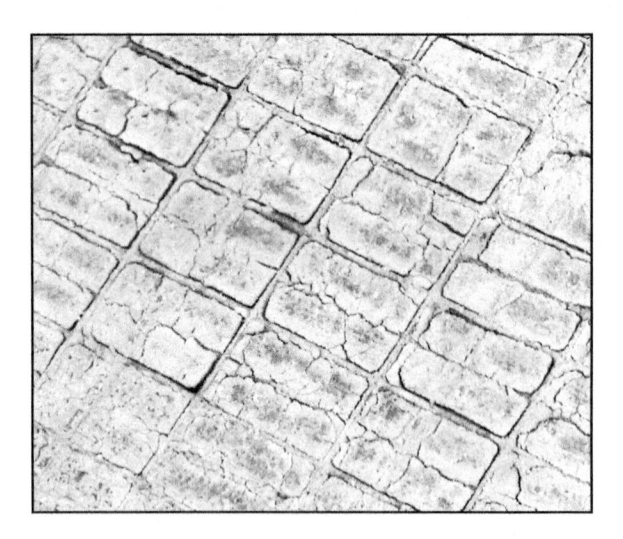

Surface detail of the oxides on a salvaged negative plate

Lay the groups on a plastic tarp. Cut the straps from the plate lugs as close to the straps as possible. You will need to burn a piece of lead to these lugs later to make up for the piece cut off that is left in the strap. Set the straps aside for melting or reuse and lay your plate grids on a piece of plastic.

Using dental tools, poke out the oxides that are contained within the grids. Start at the center of each grid space and then proceed to the edges. Do this to every grid space and scrape the grid thoroughly clean of the oxides.

Recycling Battery Parts

A thoroughly cleaned plate grid

The grid is fragile, so take care not to break the grid lines. If you do happen to break or bend a grid piece, you can usually push it gently back into place. It can be tedious work, but is worth the effort and time. The cleaner the grids, the better the plates will work in a new battery.

LEAD CASTING

Lead and lead alloys have a low melting point, so a simple propane stove and torches are all that is necessary to work with this metal. Melting and burning lead should only be done in extremely well-ventilated areas. I prefer a shed that's open on two sides.

A lot that can be said about melting, pouring, and casting lead and the art of lead burning. There is however, no substitute for practice and experience to get a first hand feel for the materials and techniques. Working with lead can be fun, interesting and quite rewarding when you cast good parts and view your final work in a finished battery. Just remember to do it safely.

Tools for Melting and Casting Lead

Stoves for melting lead

You can use one large burner propane stove for everything, or have several stoves for different functions. One stove can be used for melting while another can keep molds hot for pouring; or you can melt on one burner and pour onto a griddle plate on another burner to make plates, and so on. What you need depends on what you intend to do in the melting and pouring process.

The best practice is to pour into a hot mold. This allows the metal to cool at a slower rate, which reduces defects in the casting. So, an extra burner to keep molds hot next to the melting burner is a good idea. As an alternative, you can work with warm molds that have been heated in an oven and then brought to the melt site, but it is easier to have it all there.

You can pour into molds at ambient temperature if the ambient temperature is not too cool, but do not pour lead when it is cold outside. The closer the mold is to the temperature of the melt, the better the part that will be produced. A mold temperature of about 350°F is good for most work.

Stove gas is supplied from a 20 lb propane tank with a regulator. I usually set up stoves on a ⅛" thick aluminum plate that can be shimmed up to level. I keep a level on hand for the stoves and the work platforms for pouring and melting so that adjustments can be made continuously as needed to keep the molds level.

Tools for Melting and Casting Lead

Crucibles

You will need at least one melting pot with a handle; and a griddle for holding molds for heating and or making plates. These both should be cast iron.

A pouring lip on the melting pot makes pouring more accurate and easier. The griddle should be at least 10" in diameter, though a 12" is good to make larger size plates from pours onto the griddle.

There are also square and rectangular griddles of larger size that can be used, as long as your stove can provide even heat over the whole bottom of the griddle. If there is heat in the middle only, the corners will remain cool and cause problems in pouring and melting.

77

Before melting lead in a cast iron melting pot or on a griddle, burn off the coating if there is any so that it does not contaminate your first melt. This usually takes about an hour at moderate temperature on the gas stove. The pot or griddle will smoke a lot and smell a bit at first but when the coating is burned off, that will subside.

When melting lead in a pot, skim it every few minutes to remove the dross of oxide that forms on the surface. This can be done with a spoon (discarded kitchenware only, please); or a cast iron skimmer, as in the photo above.

You can also prevent oxides from forming on the surface of melting lead to some degree with a flux, but it's not necessary and I rarely use it. Common fluxes are borax and or beeswax. A small dime shaped piece of beeswax thrown on top of the melt will help to reduce oxidation and consolidate the dross. I prefer to simply skim the surface of the lead frequently.

Heat and splash protection

To work with pour pots, griddles and other metal implements for melting lead, you must wear heat resistant gloves. Very thick leather foundry gloves are ideal for the task. Test your gloves out ahead of time to be sure they will give you sufficient heat protection for the full amount of time you need to complete your pours. Suddenly getting that burning feeling through your gloves will make you drop the pot of molten lead, which can cause you serious bodily harm, not to mention ruining the results of your pour and making a mess.

It is also important to wear thick cotton clothing with long sleeve shirts, full face mask or goggles for eye protection and respirator. Full body coverage is necessary to protect yourself from harmful lead splashes.

Heat and metal splash resistant cloth is good to have on hand for fire resistant clothing overlays or splash blankets to protect yourself and anything else in the melting area that could be damaged by spattering molten lead. See page 170 for more info.

Splash plates, fire shields or work platforms can be made from pieces of sheet metal. For instance, you can pour onto a metal plate to form sploot seals, or other thin pieces of metal that can be easily cut into smaller pieces for remelting during projects.

Other miscellaneous useful tools

Holes cast in molded lead can be enlarged with a reamer; or the holes can be drilled and reamed after the piece is cast with a hand drill, drill press or milling machine to make pilot holes; and then the holes are reamed to the appropriate size.

Graphite powder or pure talc with no additives can be used as mold release. I use mold release occasionally, but mostly do not find it necessary because I usually destroy the molds and don't reuse them. Depending on the shape of the part, you may be able to reuse a mold quite a few times without a release. Sometimes release does not work that well, anyway.

Lead for Foundry Work

Melting new lead

For foundry work, you can use either new lead or salvaged lead. New pot lead is usually sold in 5 lb ingots, but you can also get it in the form of shot, sheet lead or lead wool. It is either unalloyed or alloyed. Unalloyed lead has a melting point of 621°F. Lead alloyed with antimony has a lower melting point.

Unalloyed lead is usually about 99% lead with incidental impurities.

Alloyed lead has another metal or metals intentionally added to it. Lead hammer metal is typically available and is about 94.25% lead and 5.75% antimony. This is a good metal for making grids for pasted plates. It can also be used for castings such as straps and posts. Ingots should be either melted down and poured into smaller pieces; or cut up into smaller pieces with a hammer and chisel. Smaller pieces of lead make it easier to weigh out the correct amount for a given melt.

5 lb lead ingot

If you are casting strap posts, for instance, and do not have an original to weigh to tell you how much lead is needed, you will have to estimate the weight for the first melt. After the first melt, you can weigh the cast part and know exactly how much lead is needed for the next melt for the same part.

Lead for Foundry Work

Whatever form of lead you start with, working with smaller chunks helps to avoid wasting a lot of gas melting much more lead than you need for that particular piece. However, always add at least one quarter more lead than you need for the melt. The molten lead gets lost as spill-offs, hardening in the pot as you pour, and from dross skimming.

Melting salvaged lead

For salvaged lead, first clean the metal well before it goes into the pot, and reduce it into smaller chunks if necessary or possible. There are likely to be oils and other impurities present so the lead may smoke as they burn off. Skim and clean the molten metal well before pouring.

If you are melting wheel weights, stir the melt as it proceeds to bring the clips to the surface. The clips are made of a lighter metal, so as the lead wheel weights melt, the clips will rise to the surface when stirred. They must be removed. This can be done with a good skimming ladle.

Above, a lead ingot chopped into smaller pieces for melting. Below, melting the chunks.

Salvaged lead wheel weights are a good source for casting lead.

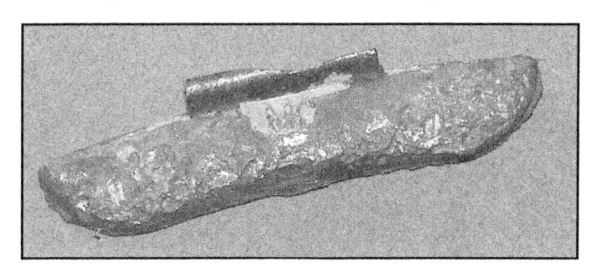

Making Molds for Lead Casting

Fabricating battery components for the most part consists of melting and pouring lead or lead alloy into forms and molds that have been impressed with a pattern. The forms or molds are made with plaster of paris. A pattern, which is the exact dimensions of the part you wish to reproduce, is inserted into the form, and wet plaster of paris is poured in. When the pattern is removed from the plaster, a void is left. When the mold is thoroughly dried (cured), molten lead is poured into it to produce the cast part.

Many types of containers can be used to make forms. Quick forms can be shaped using sheet metal such as aluminum roof flashing. Molds made from sheet metal can be reused many times. You can also use cardboard, plastic or any container that is the right size.

Recycled post and intercell connecting strips can be used as mold patterns. Above, the raw parts to be used for patterns, and below, the parts trimmed, cleaned and ready to use.

Patterns

For parts patterns, you can use the original part recycled from a battery (see right), or make a plastic or metal pattern based on the original, or make your own design.

82

Making Molds for Lead Casting

If you are using recycled battery cases, measure and make sure that any part you design will fit and mate with the case holes. For instance, if you need a battery post that is ½" diameter, use a ½" diameter rod for the pattern. Any design for a part that will go into a recycled case must fit the case and take into account the plate positioning. You may use all the measurements of the original parts exactly as they were in the case, or you may modify the patterns to make design changes. As an example, the strap patterns made for the recycled cases in this book were made shorter because the new battery would have fewer plates than the original battery had. I could have used a strap the same length as the original, but that would have been a waste of lead, adding dead weight and more resistance to the battery.

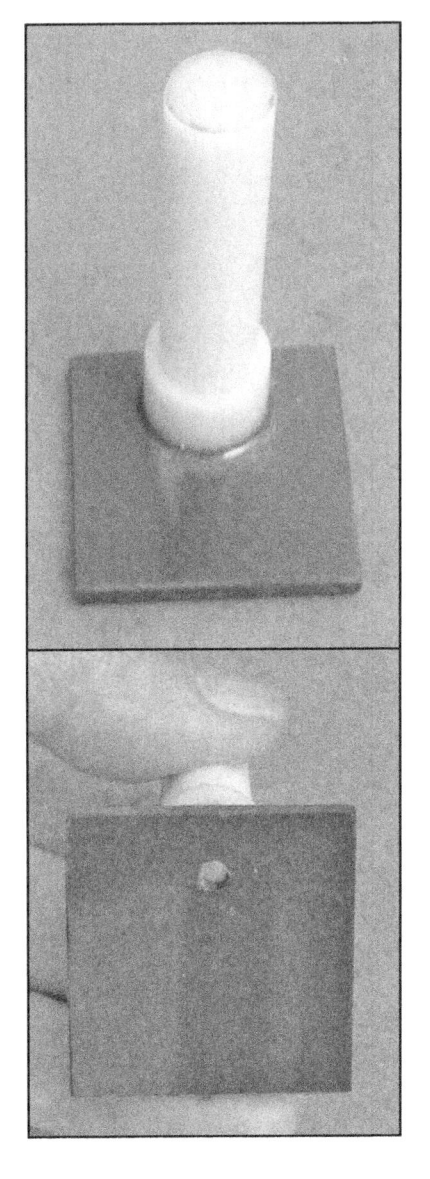

Patterns are very easy to make from plastic sheet, tube, and rod. The components of the part pattern are glued together with epoxy. Use very good epoxy as the pattern must hold up under the strain of being pulled out of the plaster of paris mold once the plaster is set.

All edges and corners of the pattern should be rounded off, all the surfaces should be smooth so that the pattern is easy to pull from the mold.

83

If the original part will be the pattern, simply clean it up, cut off any parts not needed, file it smooth and then clean with a steel brush. Original parts can be modified by burning on, or cutting or filing.

Whether you use original parts or fabricated patterns, drill small pilot holes in the base of the patterns where you will be pulling it out of the mold. Screw hooks in these holes after the mold has set so that you will have something to grip to pull the pattern from the mold.

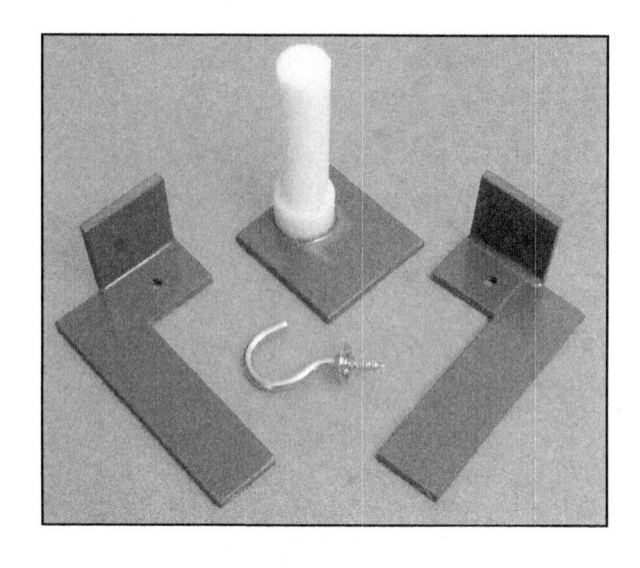

Making forms

Forms need to be large enough to contain the pattern part you are going to copy. The distance between the wall of the form and all the surfaces of the pattern cavity must be thick enough so that the plaster of paris mold will not disintegrate during use, but thin enough so that the drying time for hardening and curing the mold is not unreasonable. The distance should be at least ½" between any internal part of the cavity and the mold wall for the mold to be sturdy.

Plastic containers with snap-on lids are good for molds. Cut off the bottom of the container, snap on the lid and turn the container upside down. Then, place the pattern part into the mold and pour the plaster of paris, covering the pattern part by at least ½".

Making Molds for Lead Casting

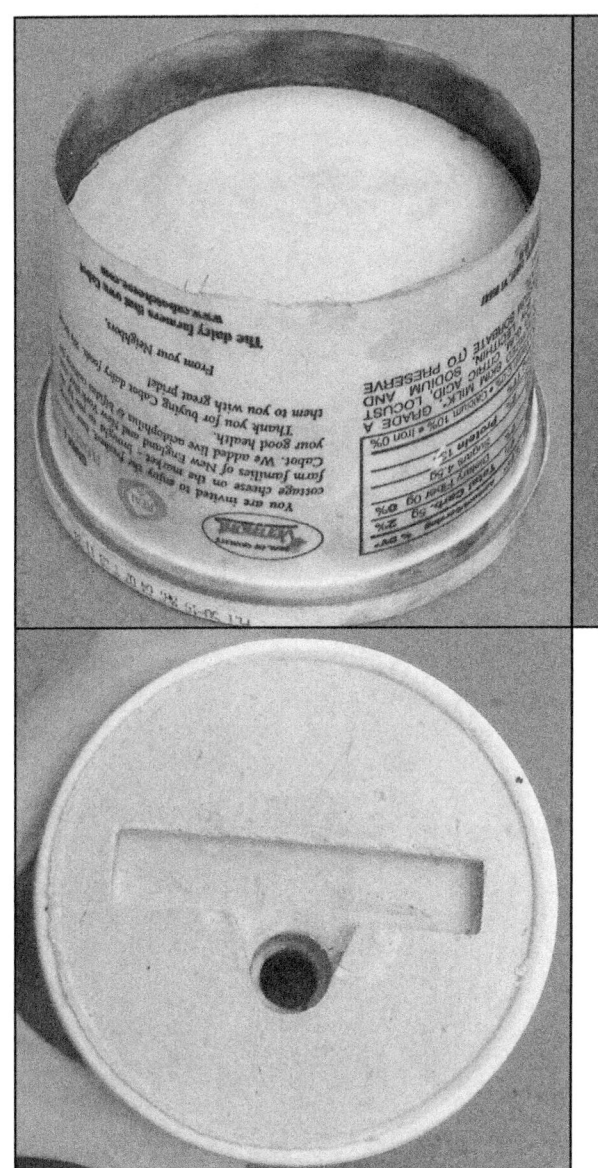

When the plaster has hardened, take the lid off and pull out the pattern.

Excellent forms can be made using aluminum flashing. This material is thin and easy to cut with metal shears. For squares or other shapes that need corner bends you can use a small metal break, or two pieces of wood and c-clamps to crease the metal. To use the form, simply put a rubber band or two, or tape, around the outside to hold it together while you pour the plaster of paris. You can also make simple round forms from flashing by rolling it up to the appropriate size, and then wrapping it with a rubber band or tape.

Cardboard flats will work, too, and can be shaped very easily and held together with paper clips, tape or rubber bands. Cardboard tubing of the correct diameter can be cut to length (see photo next page).

The Battery Builders Guide

Left, forms made from aluminum flashing, and PVC pipe.

Below, form made from a cardboard tube

Thin-walled PVC pipe can be used. Make a slit down the side so that when the plaster has hardened you can move and bend it a bit for easy removal of the plaster mold.

Many different types of objects and materials can be used to make the mold forms that you might need. The important thing to remember is to make them so that you can use as little plaster of paris as possible to minimize drying time; but thick enough to have the structural integrity needed.

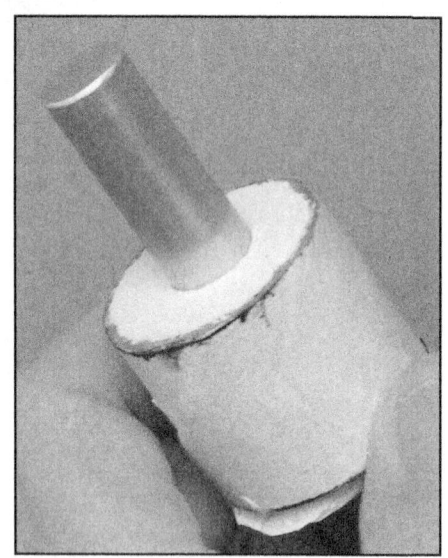

Mold platforms

Whatever type of mold you make, you will need a platform to set the form on while pouring. This can be a flat sheet of plastic or metal. The platform needs to be a leveled flat surface.

Making Molds for Lead Casting

Set the bottomless mold on flattened modelling clay, and press down like a cookie cutter

Remove the clay inside the mold, then apply vaseline to all inside surfaces, and to the pattern piece..

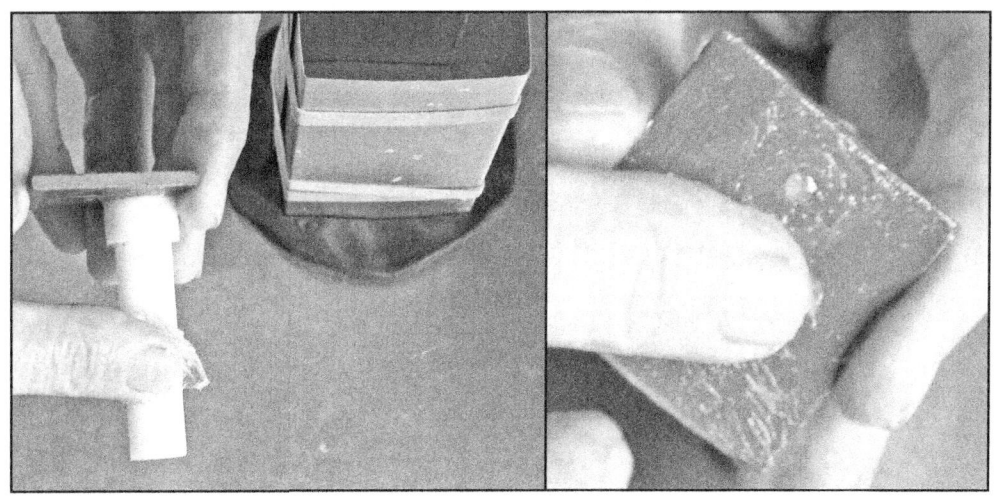

87

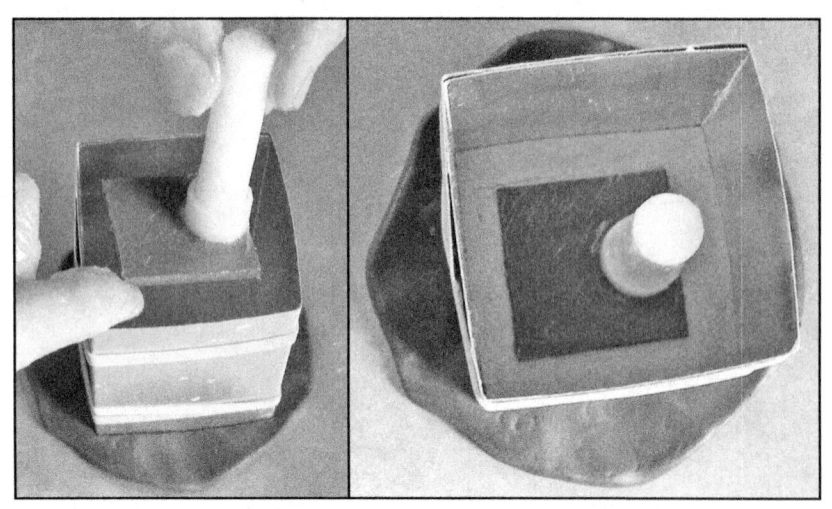

The pattern is coated and placed in the form, ready for the wet plaster to be poured in.

If the form has a bottom, such as the containers with snap-on lids, you do not need to be concerned about bleed-through of the plaster of paris from the bottom of the form.

Open-bottomed forms should be set on a perfectly flat surface. The edges can be packed with a thin cover of molding clay where the bottom of the form meets the platform. This keeps the plaster of paris from leaking out.

Lay a flat piece of modeling clay on the platform, and set the form on top of it, like a cookie cutter. Push downward until the form edges dig in and touch the surface of the platform. Then, pick out the piece from inside the form. This provides a good seal, fitted to the rim of the form.

Release agents

Patterns for use in the forms are always covered with a release agent such as Vaseline. This enables easier removal of the pattern from the mold.

Coat the patterns liberally with Vaseline (see photos), but apply the coat as smoothly as possible.

You can also lightly coat the platform inside an open-ended form (left) and the side walls of the form to aid separation of the plaster from the form wall. Dry PTFE spray or spray silicone can also be used for this, but

Making Molds for Lead Casting

Vaseline works well and is readily available and inexpensive. The plaster mixture must be poured into the mold immediately upon mixing, so before you actually mix the plaster, set up the forms and patterns for pouring so they'll be ready. Place each pattern in the center of its form.

Plaster of Paris

Although there is a variety of commercial metal casting plasters that could be used for lead molds, plaster of paris is easy to obtain at most hardware stores, works well, and is the least expensive option for molding purposes.

Do not use any substitute for plaster of paris, especially if it contains plastic compounds. Use plaster of paris or a metal casting plaster only.

Mixing plaster

Plaster of paris is easy to mix: combine one part of water to two parts of plaster.

When mixing, always add plaster to water, and do it in small increments, stirring it in thoroughly to keep it smooth. Too much plaster powder all at once will make a lumpy mixture which is no good for making molds.

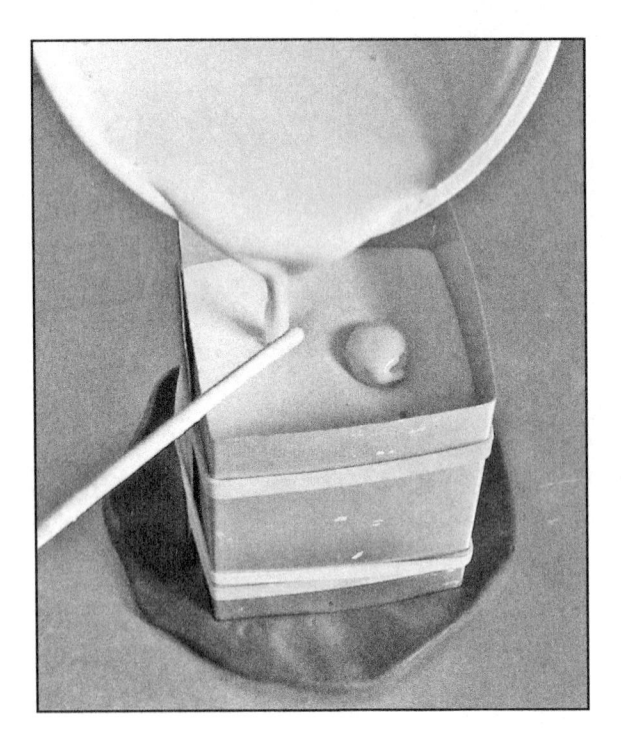

Pouring the mold

Pour slowly into the form and then use a stick to even off the plaster at the top of the form, and wipe off any excess plaster.

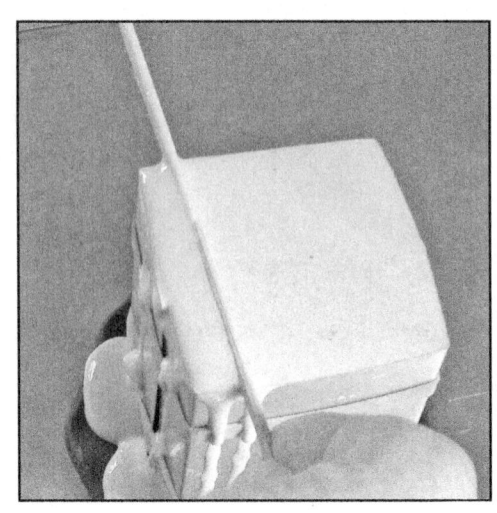

Making Molds for Lead Casting

Removing the pattern and form

The hardening time for the plaster will vary. Factors that affect this are ambient temperature, humidity, and the temperature of the water you use in mixing. Usually it will set up in 20 to 30 minutes. If you wait too long before pulling out the pattern, you may not be able to get it out without creating defects in the form. If the pattern is taken out too soon the mold will be too soft and pulling the pattern will create defects in the mold. Though this may sound problematic, in fact, with a little practice you'll get a feel for the proper time to pull the pattern.

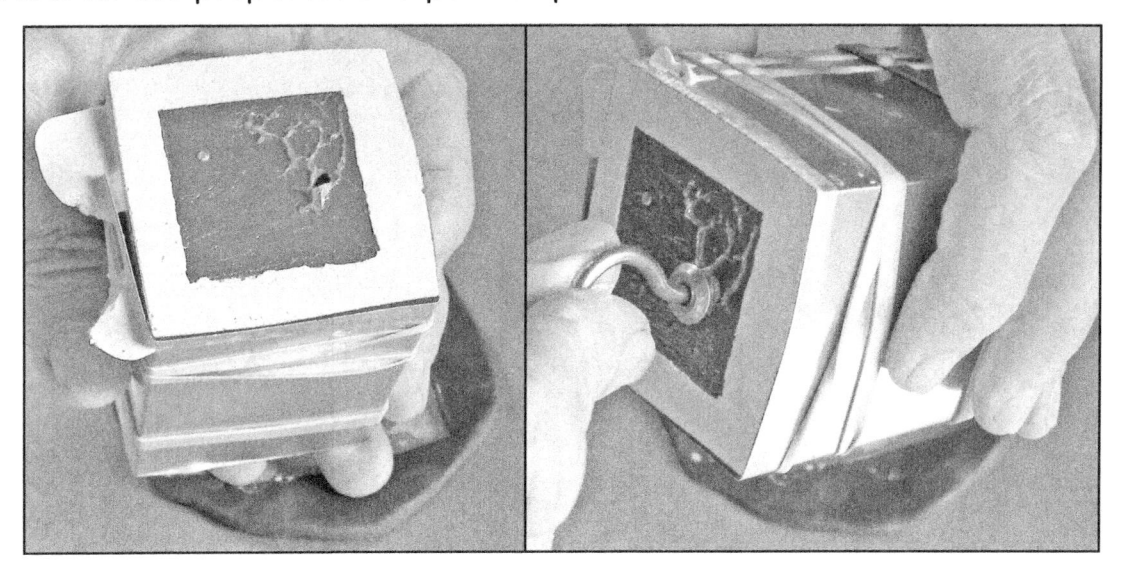

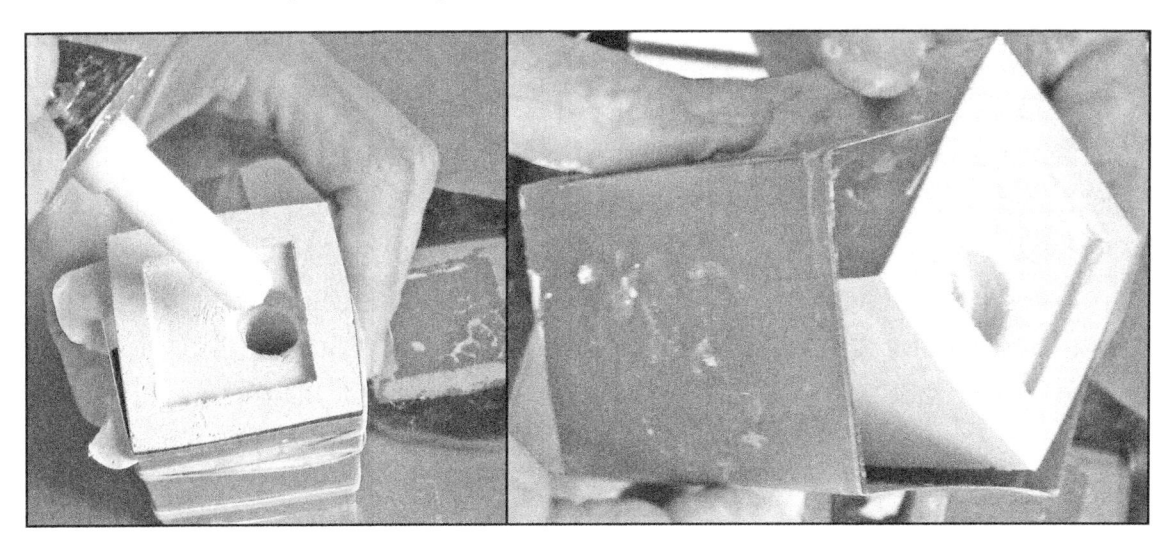

When you feel the pattern is ready to pull, turn the mold upside down, insert the hook by screwing into the pre-drilled hole. Pull the pattern from the hole with a straight backward motion to avoid nicking the edges of the mold. After the pattern has been removed from the plaster, take the form wall off.

Cleaning and curing the mold

After the pattern and form have been removed from the mold, clean out the Vaseline left in the mold with cotton swabs. Then, with a small artist's paint brush dissolve and brush out the remaining traces of Vaseline with a solvent such as xylene, alcohol or other suitable solvent. Each mold needs to be dried and cured. This takes one to two weeks depending on the method used for drying, the thickness of the mold and ambient weather conditions. An easy and inexpensive way to cure molds is to place them outside in the sun and wind to evaporate the excess moisture from the molds, if your weather is cooperative, in other words, warm and dry. If there is any chance of rain do not leave molds outdoors as they can be ruined by rain.

Making Molds for Lead Casting

A more reliable method is to place them in a conventional kitchen oven. If you have a gas oven with a pilot light, the pilot by itself produces enough heat to dry out the molds.

You can also use the oven to supply a higher constant temperature of 110° to 120°F during the curing phase. For the best curing results, stay within this temperature range. You can occasionally boost that temperature up to about 140°F for several hours a day to hasten the process. The molds should not dry out too fast as they can develop cracks as the moisture seeks a release, and the mold can also tend to get crumbly. If you do not have an oven you can use and are not in a warm dry season to cure molds outdoors, you will have to settle for room temperature curing which will take a little longer.

When drying molds, it is helpful to have them laid out on a grid or grill rather than a solid surface, so that air can flow across the bottom of the molds and dry them out.

When fully cured, a mold will make a higher pitched tinny sound when tapped with your fingernail. The mold will also feel dry to the touch, and weigh noticeable less due to the shedding of water.

Be sure the molds are well dried before use. Never use a wet or damp mold because when hot lead is poured into it, it can explode due the fast formation of steam from the water in the mold coming into contact with the very hot lead.

Before starting a specific project it would be advisable to make a few molds for practice to get a feel for the process. Always make more than you will need so that if something happens to one or several molds you will have spares ready to replace them. This will save you from having to stop your project and wait for a new batch of replacement molds to cure.

Pouring the Lead

Practice pouring

Before actually doing any lead pouring projects, practice pouring to get the feel of it. Notice the characteristics and time frames for melting different shapes and sizes of lead pieces.

When melting for a pour, keep the lead on the heat about four or five minutes after all the pieces have fully melted and coalesced before you pour. When you pour, the lead should be neither too hot nor too cold. If it is too cool, it will harden too quickly as it hits the mold. You will get a good idea of timing after you pour a few practice molds.

Skim the molten lead frequently to remove oxide discoloration and dross before you pour.

Each mold must be cast in one continuous pour. Any hesitation can produce a poor cast. Only a few seconds pause will allow the first part of the pour to cool enough so that the later part of the pour will not bond well it.

It does not take much to melt lead on a stove burner so a medium size flame generally works well.

For each mold the timing may be different as it will vary with the mold depth and the lead that has to be spread.

Never melt and pour outdoors when there is any precipitation in the air. Any moisture coming into contact with hot lead can cause a lead explosion. Always keep molds and pots and any other pouring surfaces and utensils bone dry.

Using flux

If you wish, you can use a flux to reduce oxides. Use a very small piece of beeswax about the size of a dime or less, or a pinch of borax.

Pouring the Lead

If you use borax, be sure to dry the borax in an oven first to drive out the moisture. Borax is hygroscopic and you do not want any moisture going into your molten lead.

After dropping the flux into the molten lead, skim off the dross.

Post straps and intercell connector straps can either be cast as solid pieces without slots for the plates, or the slots can be cast in place on the strap. If the slots are not cast into the piece, they have to be sawed or cut in later. Either way, the slots must be a uniform distance from each other and must be wide enough to accommodate the thickness of the plate that will be inserted into the slot. Each slot also needs to be the exact width of the plate tab.

Casting straps without slots

To cast solid intercell connects and battery posts, warm the molds, melt the lead, and pour. Always pour into the deepest part of the mold first (see above) and pour in one continuous pour for each mold. This is very important.

Casting straps with slots

If the slots will be cast into the straps, you will need to build the plate burning rack described beginning on page 160. For casting slots, use strips of aluminum the width and thickness of the tabs of the plates. They are held in alignment by the combs of the plate burning rack with the bottoms of the aluminum strips resting in the molds to produce the slots in the strap when the lead is poured (see photos pages 98-99).

To cast the slots, the slot forms must be centered on the molds so that the slots and plates will fit together correctly when the pairs of groups are assembled into elements.

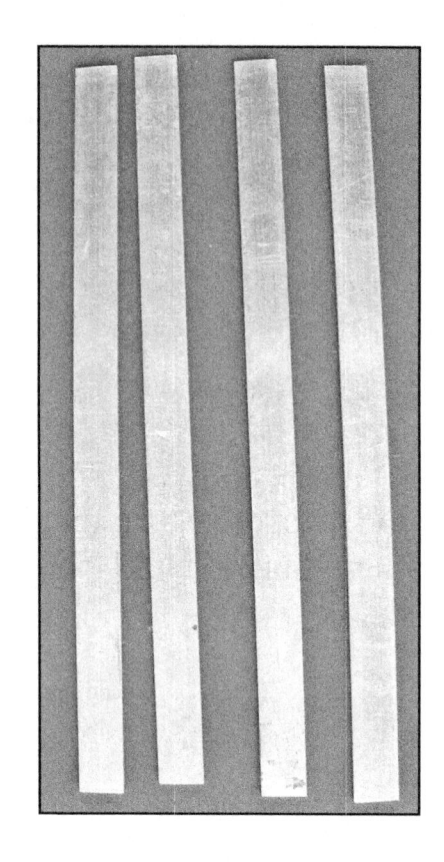

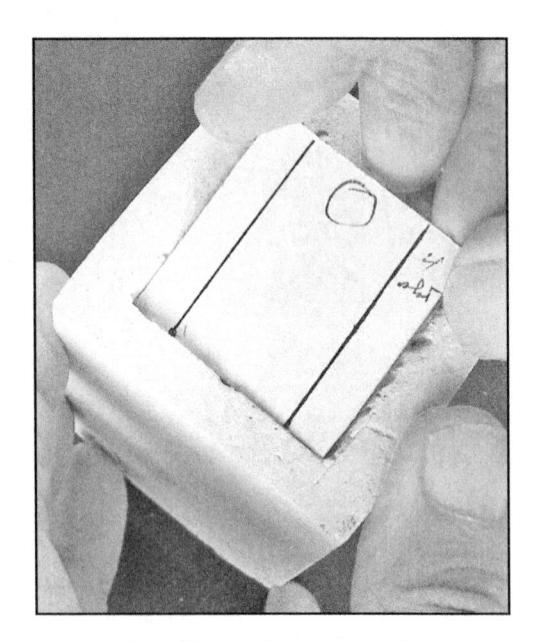

Cardboard pattern for slots in strap.

When the mold patterns were designed, each strap length was designed for the number of plates that it would hold and how thick the plate tabs would be. Use cardboard patterns placed on the molds as guides for the placement of the slots. These cardboard patterns can be drawn using the plastic mold patterns. They should be cut to fit exactly into the strap area of the mold. Then, each mold can be positioned on the plate burning rack so that the aluminum strips will be properly aligned to make the slots. For a 6 volt battery with just post connectors (external intercell

Pouring the Lead

connections), only two cardboard patterns and six molds are needed. If this battery were to have seven plates in each cell, there would be one group of three plates (positive) and one group of four plates (negative) in each cell. One pattern would have the space marked out for three slots and the other pattern will have space marked out for four slots.

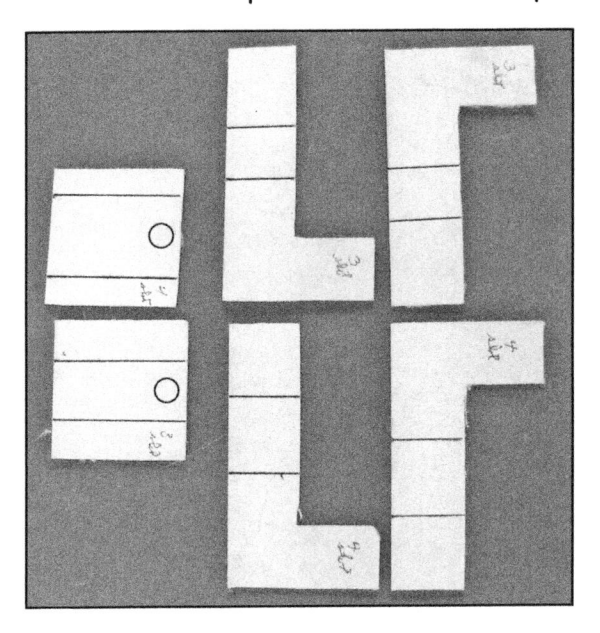

For a 6 volt battery with internal intercell connectors, you would need six cardboard patterns, as in the photo at right. One pattern would be for a positive post strap, another would be for a negative post strap, two for left hand positive and negative and two for right hand positive and negative straps. These cardboard patterns will help to place and center the mold under the strips.

Rather than mark each slot on the cardboard pattern, I simply mark the outside position of the end strips on the pattern. The combs of the burning rack are already cut and spaced, so the other slots will be aligned automatically.

Aligning the mold

To prepare a mold for casting with straps, first set up and level the burning rack. Place the rack's combs on the threaded rods as shown in the photo below to stabilize the inserted strips.

Place the strips into the comb slots. For this example of a seven plate element, for a positive strap there would be three strips. For a negative strap, there would be four strips. Hold the strips in place with one hand, and place the mold under the strips.

The Battery Builders Guide

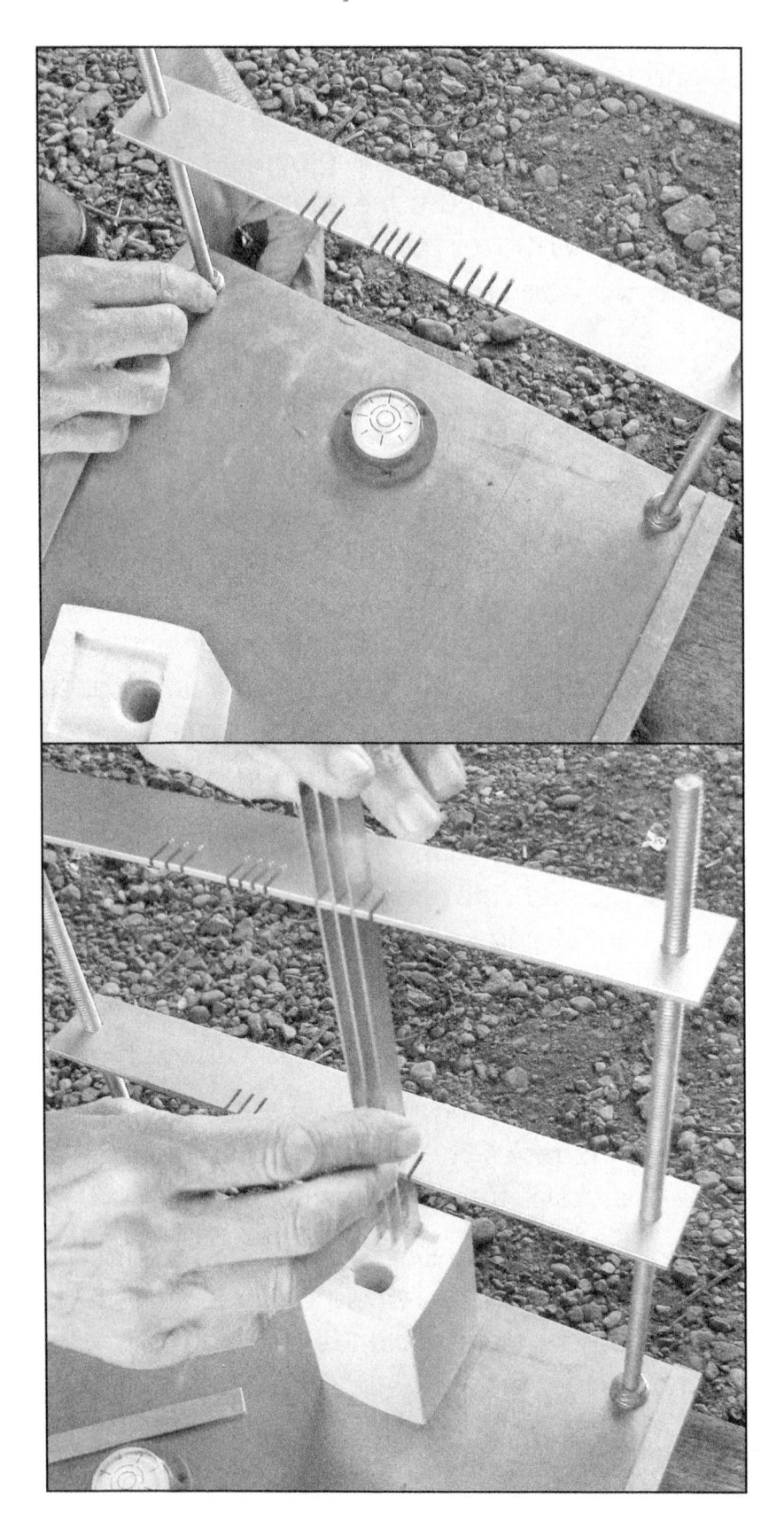

Pouring the Lead

Then, put a knob of modeling clay on the top of the strips as shown in the photograph. Press down firmly to set the clay around the strips. This will help align all the strips at the same time as you position the strips on the mold. You can lift the knob of clay and all the strips at once to get the final alignment with the cardboard patterns.

Once you have the strips completely aligned and square with the edge of the mold, to further stabilize them, add another piece of modeling clay on the strips at the upper comb as shown in the third photo below.

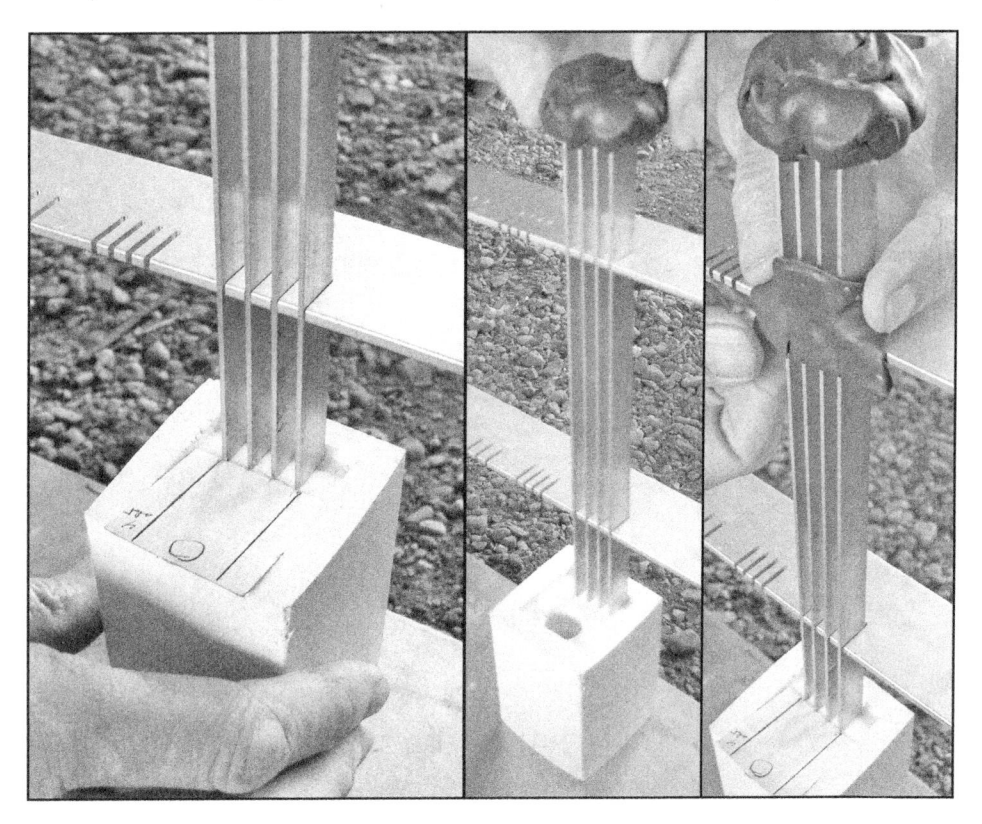

You are now ready to melt and pour the lead.

Melt and pour

Before you pour, double check to make sure the slot strips are aligned and square with the edge of the mold.

When everything is set you can begin your melt. Skim the dross frequently during the melt, then pour. Let the casting cool completely, then,

The Battery Builders Guide

move the mold and strips from the rack combs, and remove the strips from the casting.

Post castings are usually easy to remove from the mold. If you are patient, you can jiggle them out of the mold and use the same mold over again and again. If the casting sticks to the mold, you have to break the mold. It's best to assume that you will have to break every mold after one use, although it is not always the case.

Finishing Castings

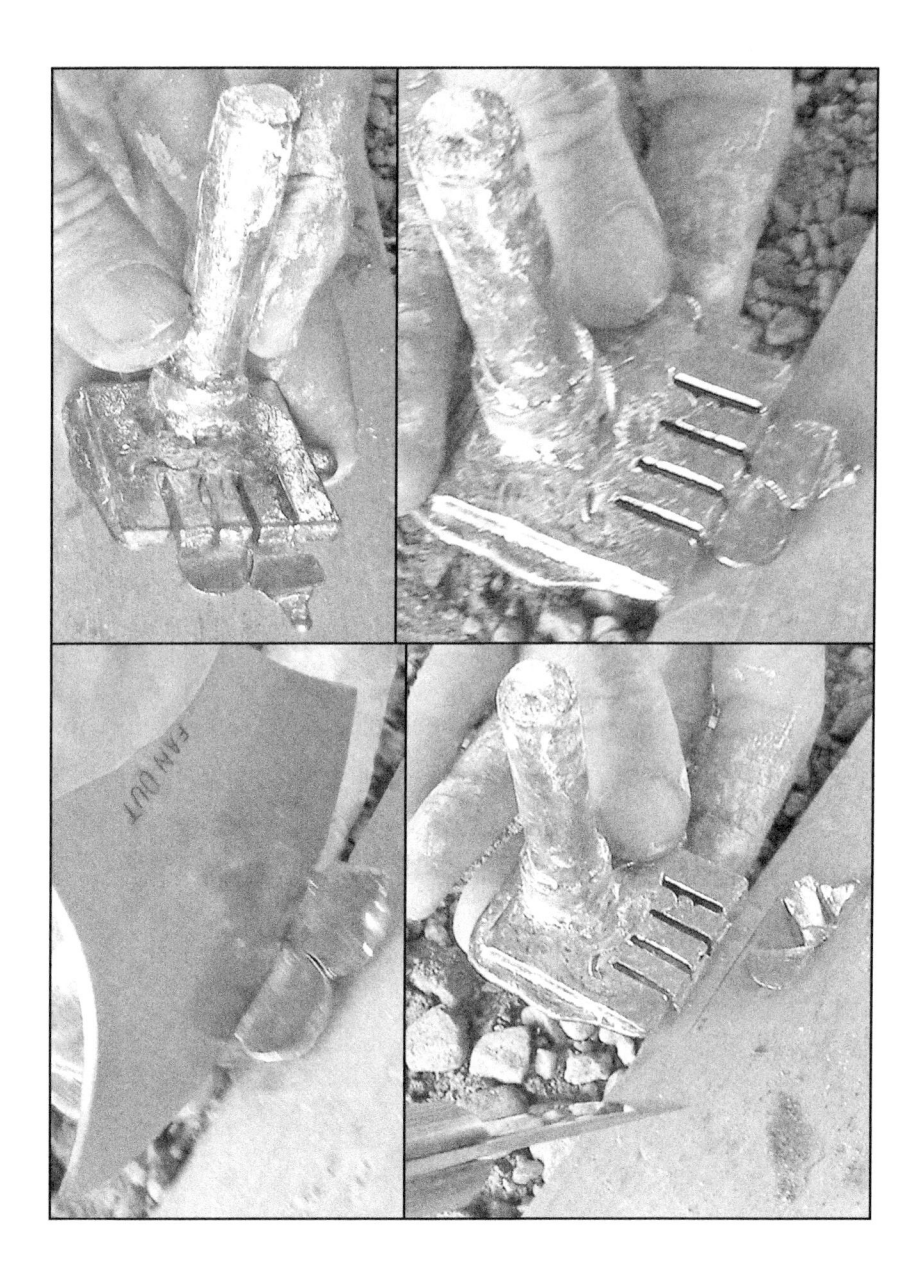

When the cast is removed from the mold, there may be defects such as overflows that have to be cut off. Use the fanout knife to cut them off and even the edges of the casting.

Wash all plaster of paris from the cast and finish any rough edges with a file, steel wool, sandpaper, or Scotch-Brite® pad to your satisfaction.

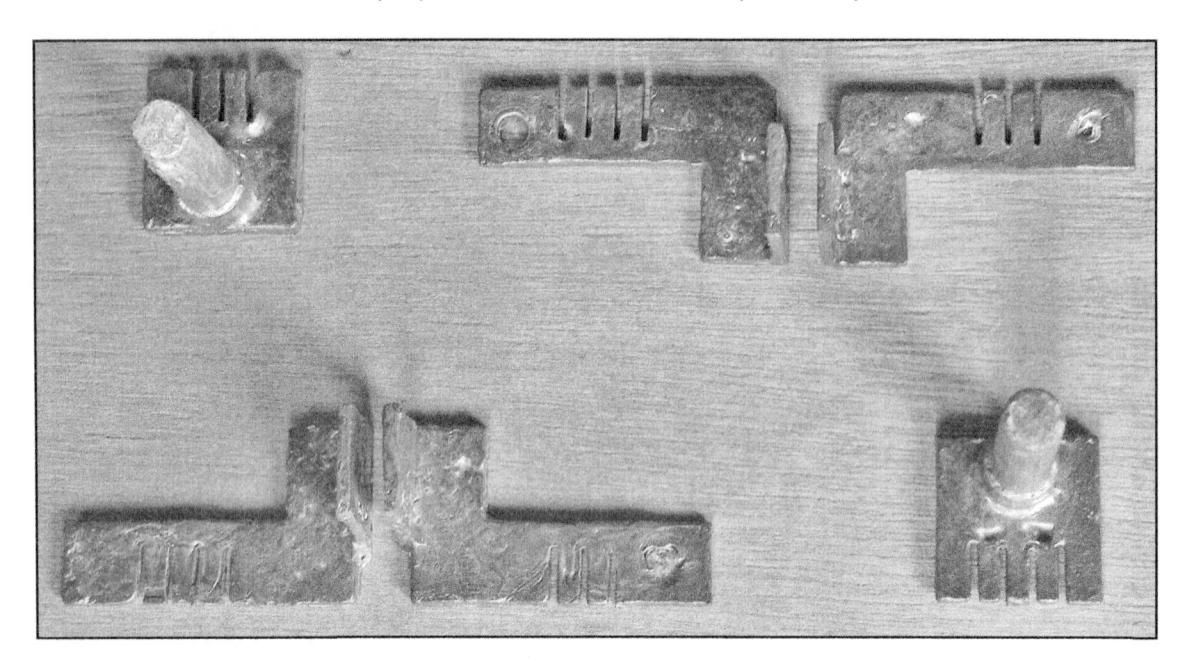

For internal intercell connectors, drill a hole in the strap where the connector abuts the wall of the cell and connects with the next intercell connector. Measure to find the center, mark with a scriber, drill a pilot hole and then ream the hole to its final size.

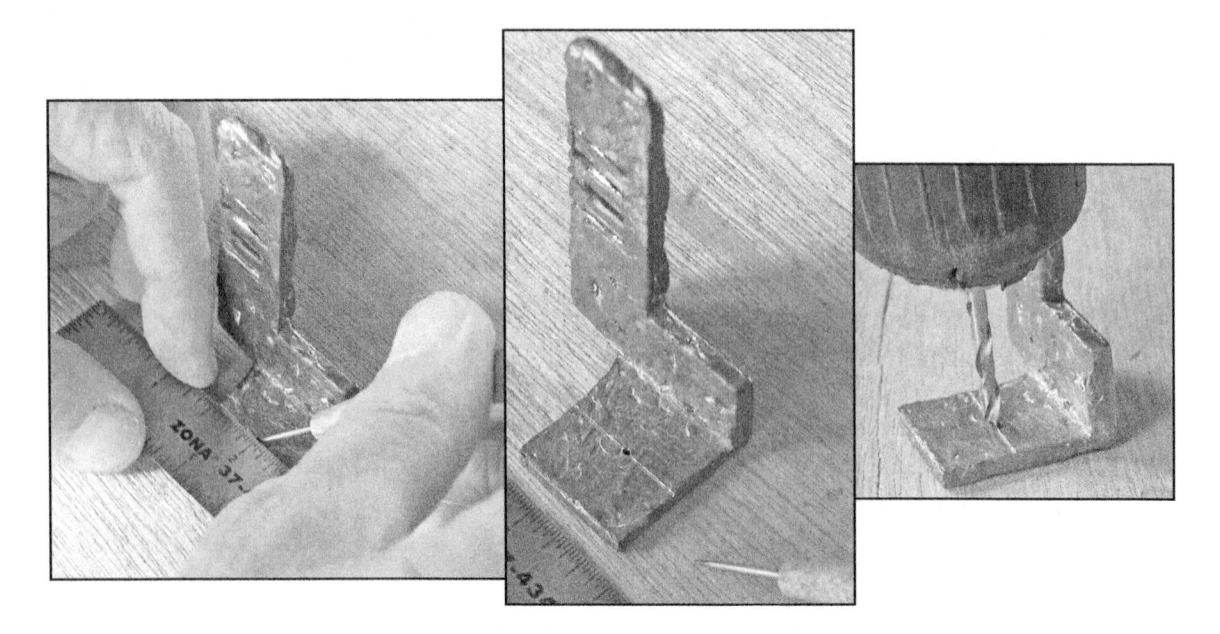

Finishing Castings

Use a reamer to enlarge the pilot hole to its final size and finish it.

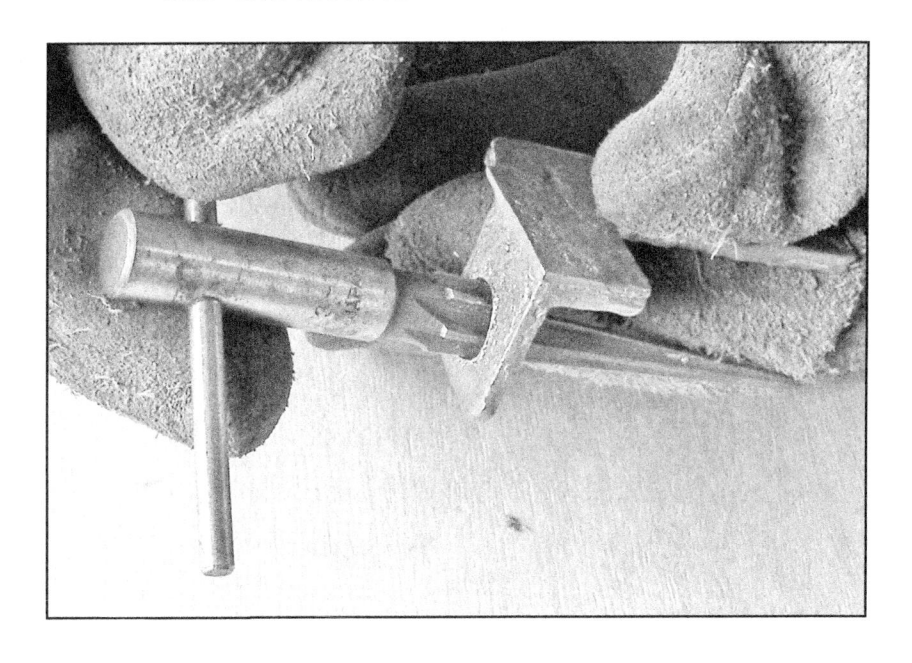

Casting Connecting Rods and Seals

Connecting rods

Internal intercell connectors are connected to each other with either a short piece of rod which is burned into the straps; or a short piece of threaded rod that is attached and kept secure with a pocket or other type of lead nut. Either plain or threaded, the rods are made of lead.

Plain rods are easy to cast. Make a mold for the diameter needed to fit the intercell hole in the case and strap. For a threaded rod, you will also have to cast nuts and thread the cast rod.

For the threaded rods used in this project, I used one of the post molds to cast a post, since they were the same size as the case intercell holes. I threaded the post part and after threading, cut off the strap part. With the strap still attached, you can grip the piece easily in a vise or in the hand for threading the rod.

A hand die holder and die (see photos below) was used to thread the rod. Since the holes and rods were ½", I used a die that would cut a 10/24 thread on ½" diameter rod stock.

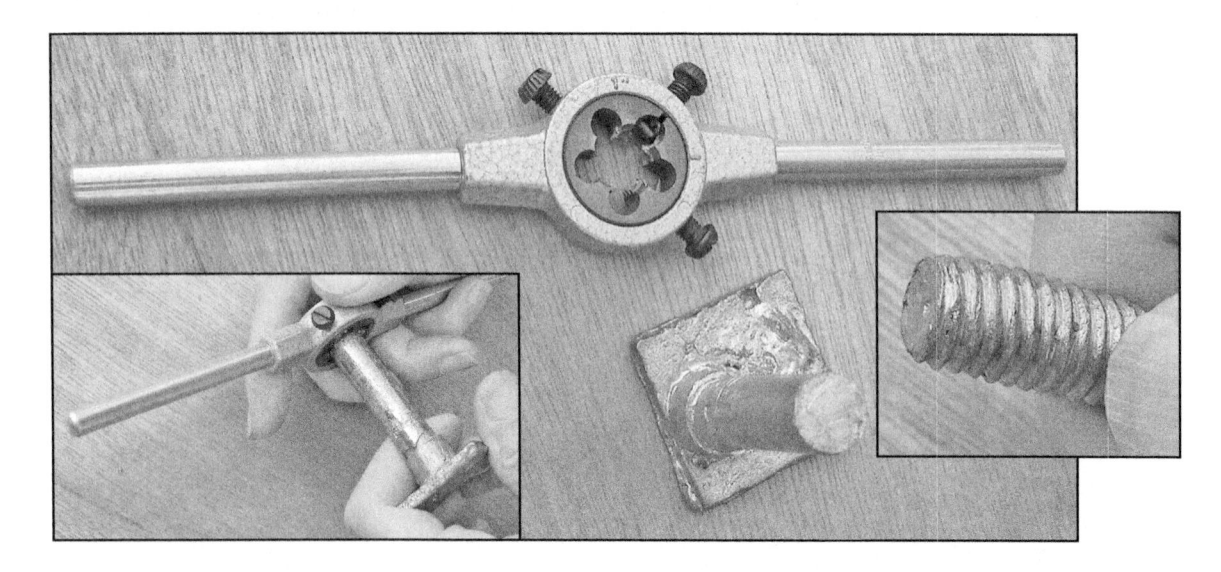

Casting Connecting Rods and Seals

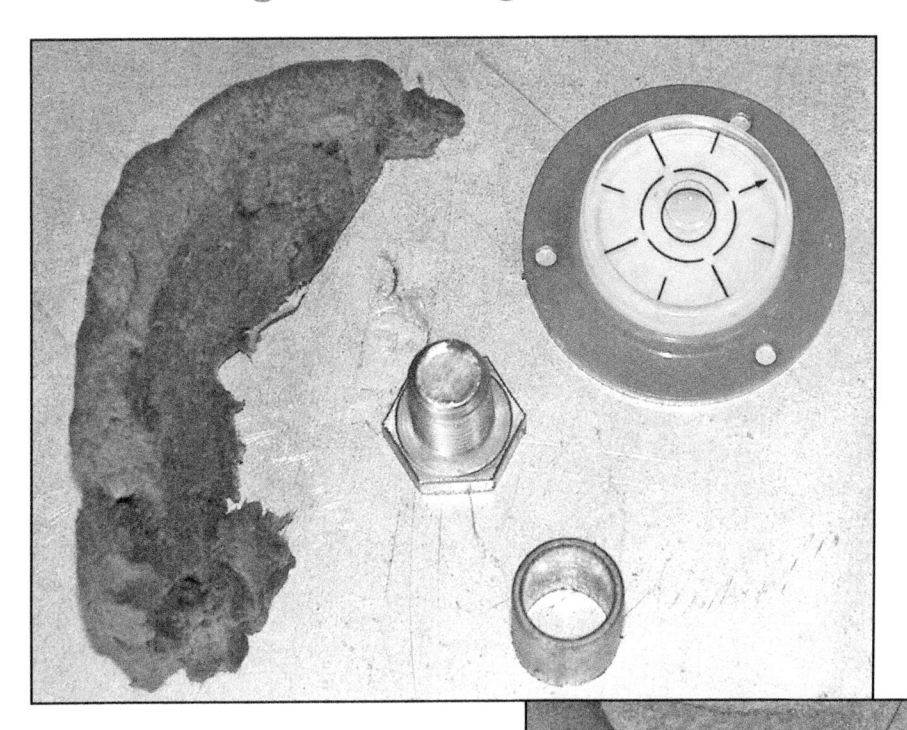

Casting pocket nuts

For casting the pocket nuts I used a 10/24 hex head screw that was ½" in diameter and 1" long in the thread area. I also used a 1⅛" long, ½" inner diameter, 1/16" wall thickness sleeve bearing. Both of these items are available at most local hardware stores. I also used casting putty.

Insert the hex head screw into the sleeve bearing and place on a level casting platform (the plate burning rack works well for this).

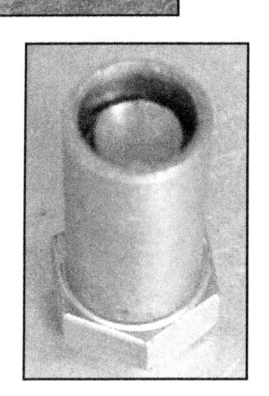

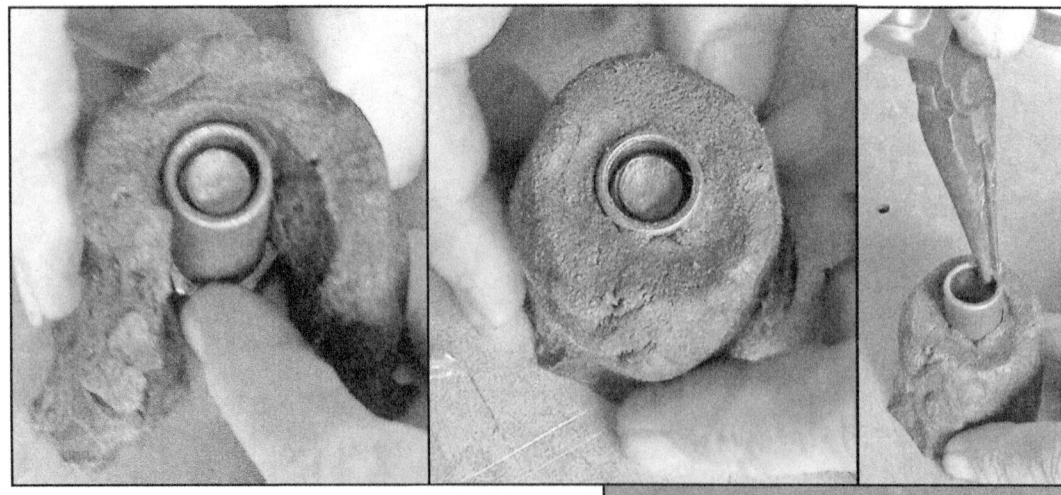

Wrap a piece of casting putty around the bearing making sure to get it packed well against the bottom hex head. Cover the outside of the bearing to the top of the bearing. The bearing is a bit longer than the screw so that when you pour your lead a cap will form over the end of the screw.

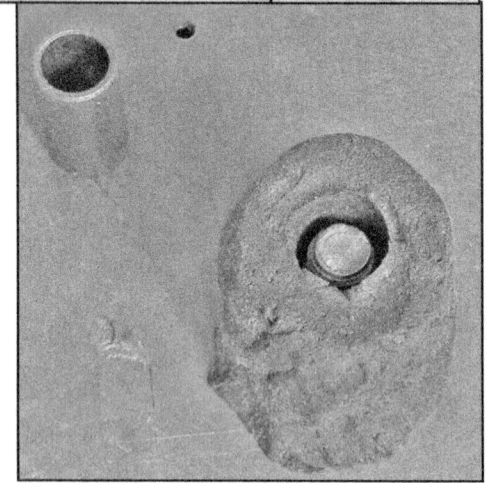

Pull the bearing from the putty mold straight upwards so as not to deform the mold. This will leave a mold cavity ready for pouring.

Melt and pour the lead to the top of the mold covering the top of the screw.

Let the mold cool and then remove the putty from the cast.

Casting Connecting Rods and Seals

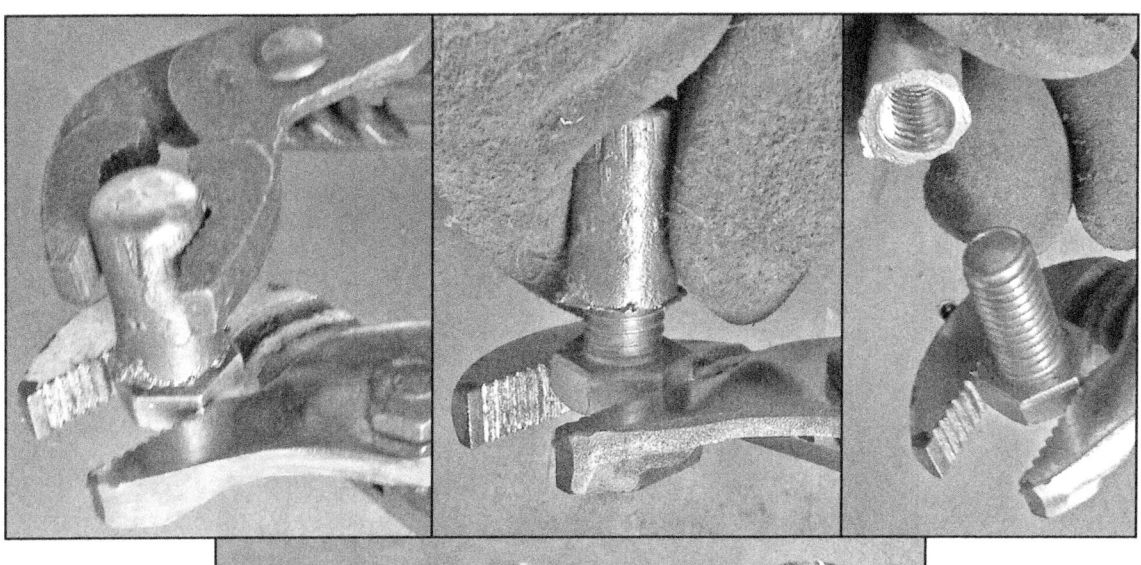

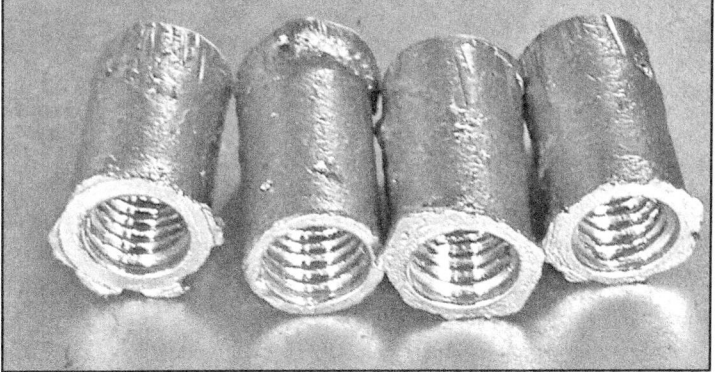

Using one or two pliers you can then twist the pocket nut from the screw pattern.

Inspect your nut and trim any excess edge, and round and file where needed.

You can use the casting putty over and over again to cast all the nuts you need for your battery project.

Casting sploot seals

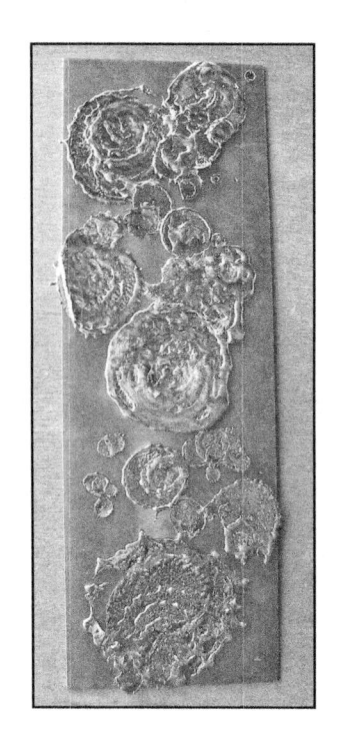

Sploot seals are foil-thin pieces of lead that can be used to overlay threads to tighten the fit between the nut and threaded rod if needed. Sploots can also be used for thin washers for a variety of battery uses. For thick washers, make a mold of casting putty, or plaster of paris, and use a washer as a pattern.

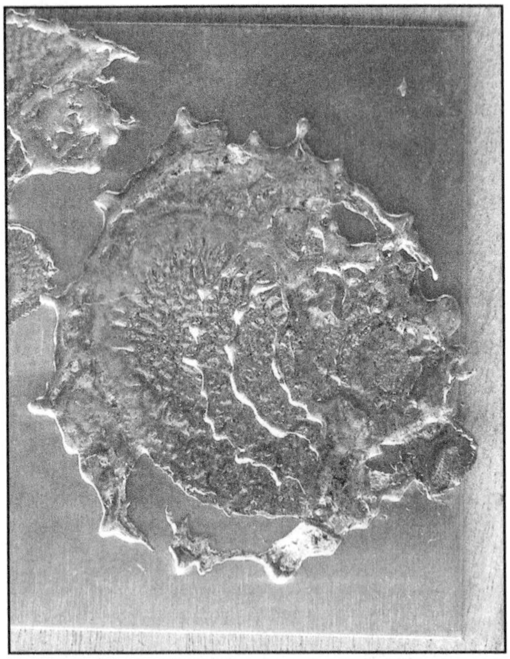

Sploots are simple to make. Melt some lead in a pot and drip it in small amounts on a metal plate. Let the sploot cool on the plate and you will have a series of foil thin pieces of lead that can be cut into washers. With a little practice you will be able to control the thickness of the sploots.

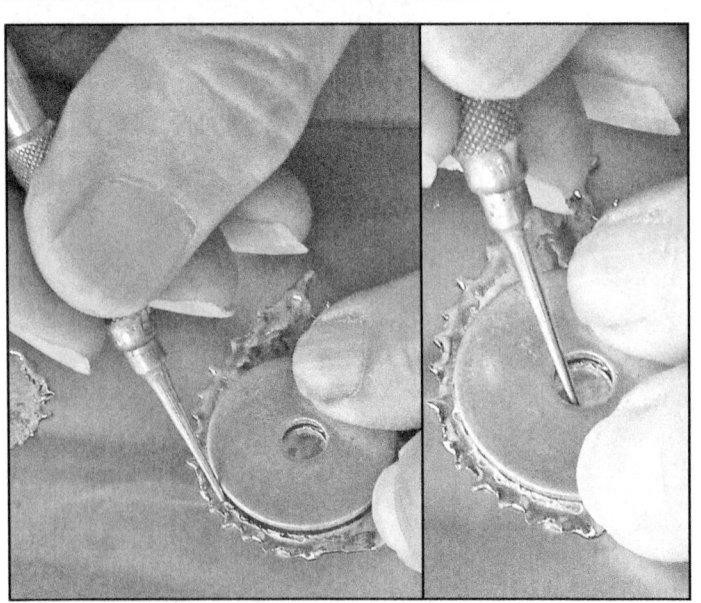

You can cut out a washer directly from the sploot using a washer as a template. Use a hole punch to cut the center hole (see photo, next page).

Casting Connecting Rods and Seals

For a smoother surface on a washer, hammer it with a mallet. You can also make thicker pours and then hammer the pieces down to the thickness needed. When you hammer the lead, check for brittleness and cracking of the metal. If it has reached that point, reheat it on a griddle just to the melting point and let it cool. This will take the brittleness out of the piece and make it more pliable, which improves its ability to provide a good seal.

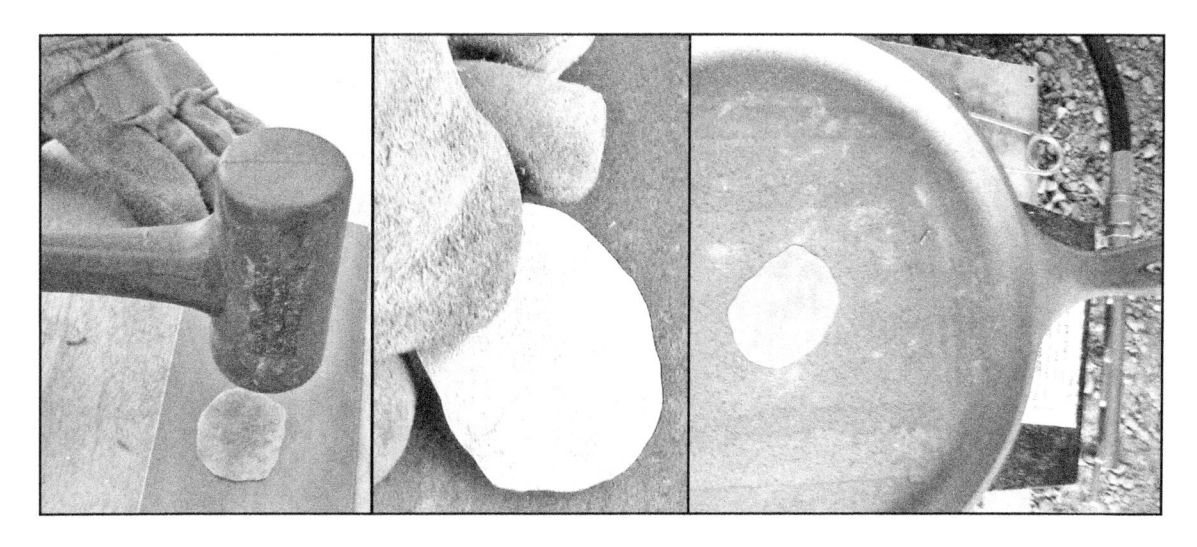

The Battery Builders Guide

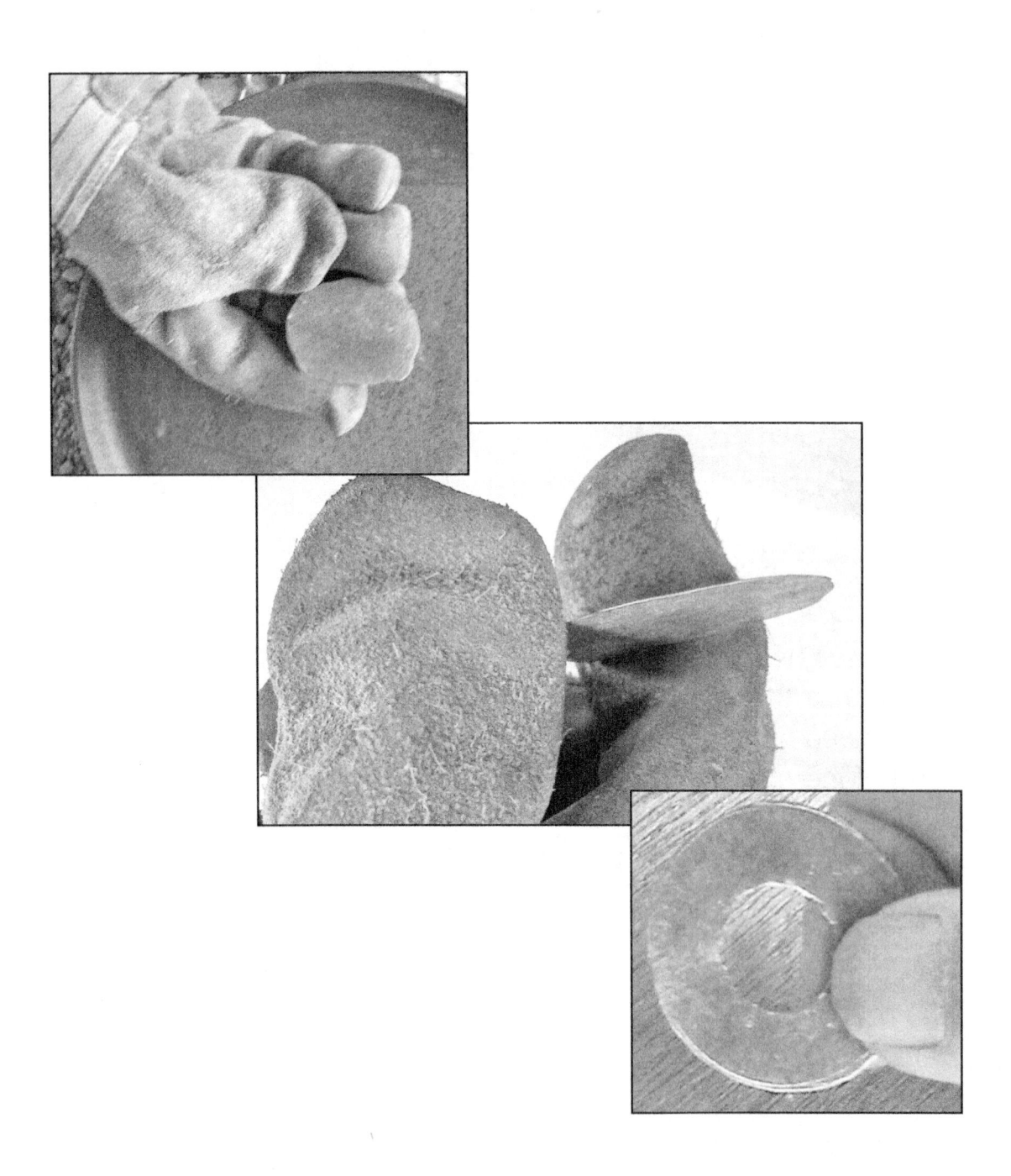

Casting Plates

Methods for casting plates

Casting decent lead plates of a uniform size is easy enough with practice, but there is an art to it, as with other lead burning and casting activities.

There are several different ways to cast plates. Grids for pasted plates can be cast using molds made from either recycled grids, or from a grid fabricated from sheet lead with the grids cut or stamped out, see page 146.

Grids and Plante plates can also be made directly from what is essentially lead sheet that you cast yourself. (Again, plate grids made from sheet lead will have to have grids cut or stamped out.)

The two molds discussed here cast solid rectangular plates. The size of the cast rectangle can either be for the exact size of the final plate without the lug extension included, in which case an additional piece of lead for the lug will be burned on after the plate is cast; or the rectangle can be sized so that the final plate, including the lug, can be cut in one piece from the cast plate. Molds can also be made to cast the exact size and shape of the plate with its lug, but this is a bit more complicated to make than a simple rectangular mold.

You will need a large one burner propane stove that will easily heat a 10" or 12" skillet, or that will adequately heat the plate mold used. You may also need an iron pot for melting lead depending on what method you use to cast the plates.

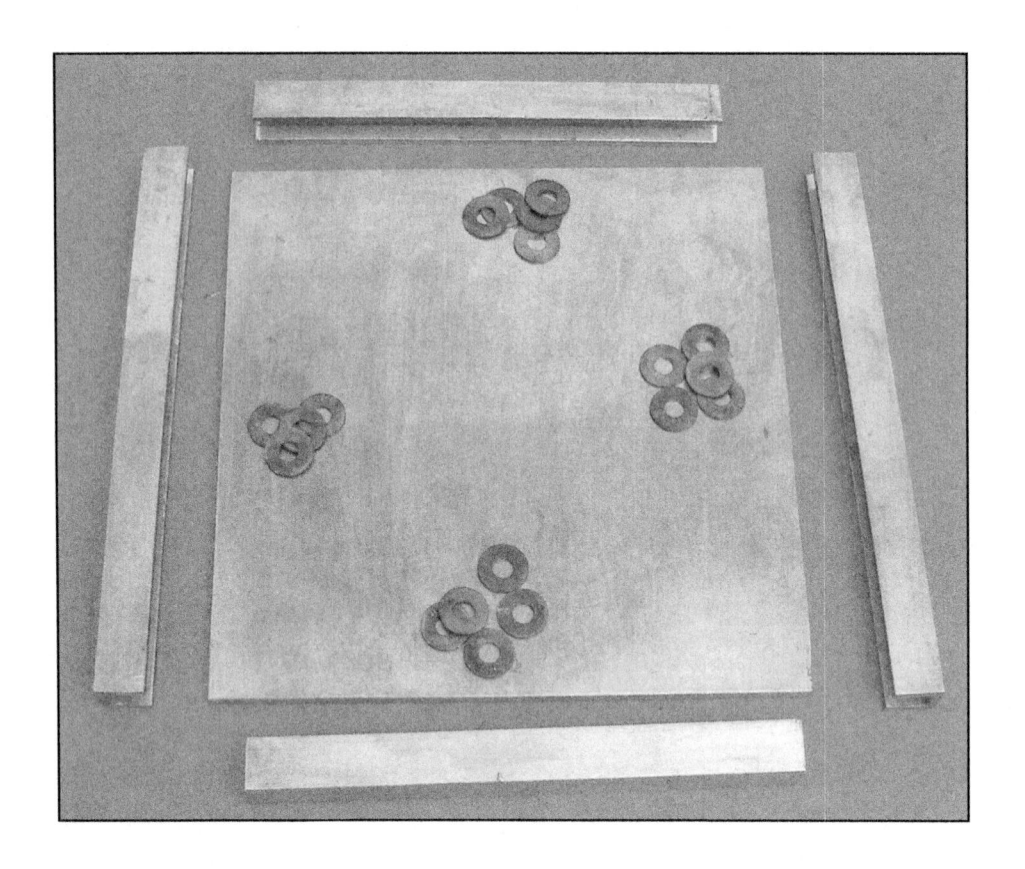

Making a plate mold

A simple mold can be made from 1/8" thick aluminum or steel sheet, U channel and some washers. The sheet should be as smooth and polished as possible on the side that will be in contact with the lead. The smoother the surface, the less it will grip the lead.

Cut the metal sheet to the size of the plate being made, with an added margin on all four sides that will fit inside the channel.

Channel comes in a variety of thicknesses. The length of the channel legs used here is 1/2". So, to make a 6"x 6" plate, the sheet aluminum was cut to 7"x 7" due to the overlap of 1/2" on each side by the channel legs.

The thickness of the plate you cast will be determined by the thickness of the channel legs. The thickness of the U channel legs in the photo is 1/16", so the plates made in this mold will be 1/16".

Casting Plates

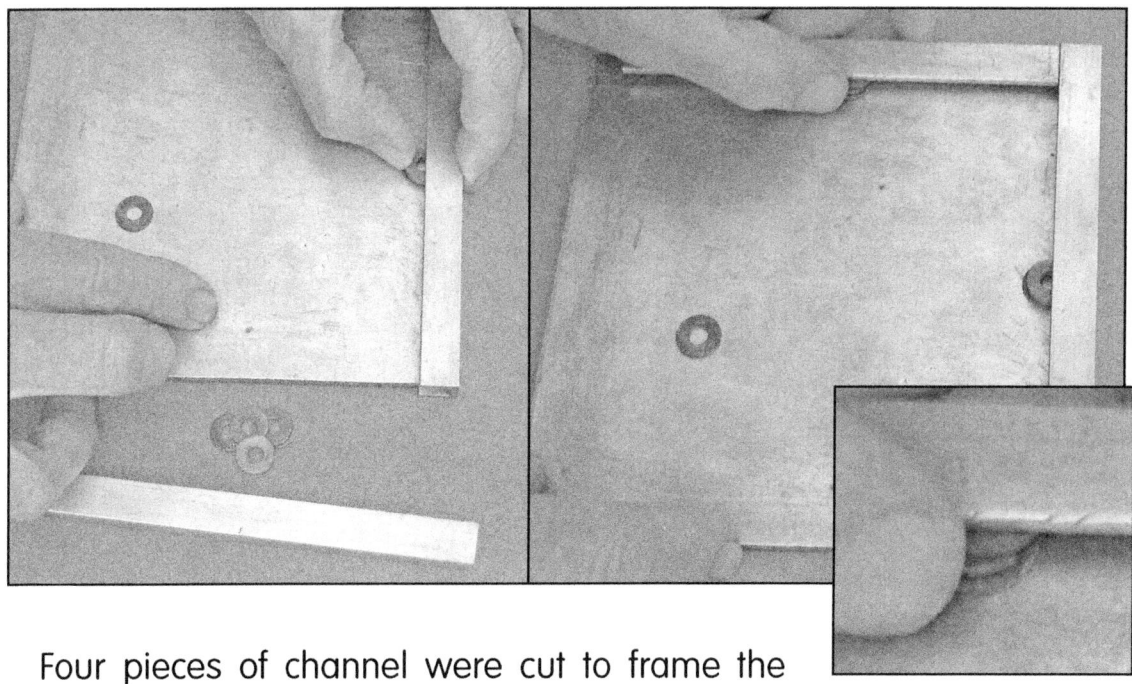

Four pieces of channel were cut to frame the aluminum sheet. Washers are used as temporary holds on the back of the mold to keep the channel frame in place until the lead has solidified. You can use anything that is metal as a temporary hold, as long as it is easy to push in and easy to remove from the underside of the mold.

The assembled plate mold. Left, the back with washers showing; right, the side that will be in contact with the lead.

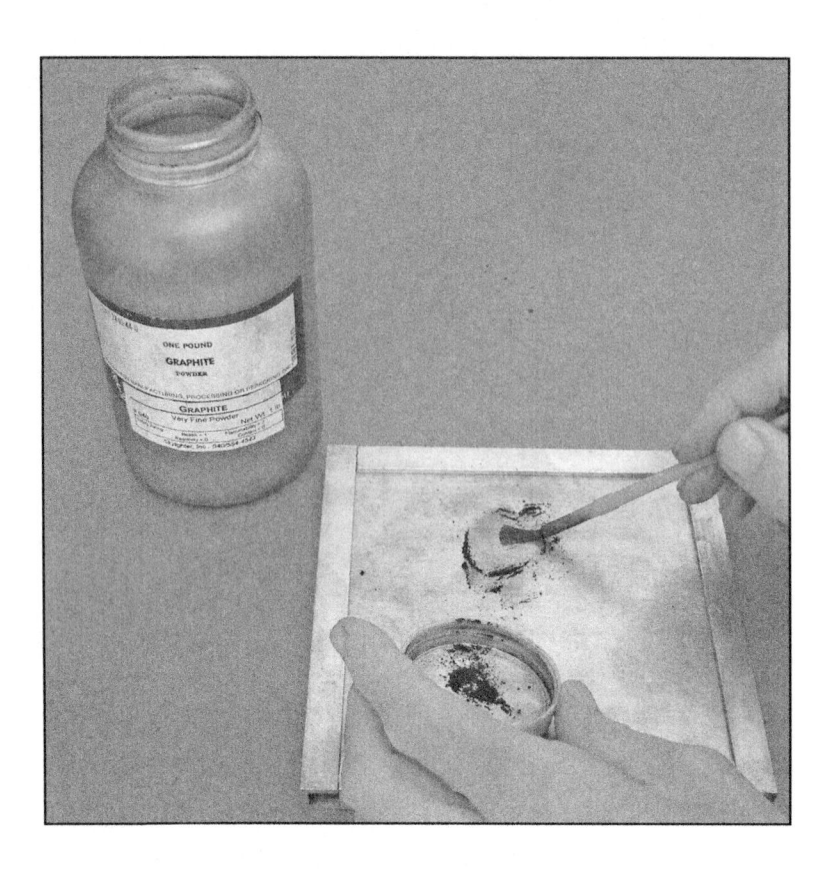

Once the lead pour has solidified, the edge pieces can be removed if necessary to release the lead plate from the sheet mold.

Release agent

Before you place the mold on the stove you can apply powdered graphite to the form to help reduce the tendency of the cast plate to stick to the mold. This is not absolutely necessary. If you do apply the graphite, rub it in well and then wipe all the powder from the surfaces that will be in contact with the lead.

Melting the lead

If it will fit securely, the mold can be placed directly on the burner of the stove, but a better option is to place it on a metal plate or on a skillet atop the burner. The lead can be melted in the mold; or in a melting pot on a separate burner, then poured into the hot mold. A hot mold can help to get a better pour and a smoother surface so that there will be less work

Casting Plates

to finish the plate. Of course if you place the lead onto the plate mold to melt, you will be doing both functions at the same time.

Measuring lead for the melt

If you are going to melt the lead in the mold, you will need to estimate the quantity of lead needed for one plate. If you underestimate, you will need to add lead pieces during the melt; and if you overestimate you will need to screed the excess molten lead from the mold form. You can calculate and weigh out the amount of lead for each mold form using the tables for weights of varying thickness of lead sheet. This will get you in the ballpark, but it is fairly easy to estimate by eyeballing after you practice your melts a bit. Also, once you have cast a plate that is just as you want it, you can weigh it to get the exact weight of lead for any subsequent plates of the same size.

If the lead is melted in a pot and poured into the mold, melt more lead than you need for a plate since during the pour some of the lead will harden on the pour spout side of the pot.

Level the stove

The stove, mold, plate and/or griddle must be leveled very exactly so that the plate you are about to pour will be a uniform thickness throughout. After placing the form on the stove and leveling everything, the lead can be melted.

If you are using a separate melting pot, start heating the mold as soon as you start to melt the lead so that the mold is hot when you pour.

When the lead has totally melted in the mold, either screed the surface to even it; or leave it as is. Turn the gas off to let it harden and cool down.

Once the plate is cool, remove the plate from the mold.

If the mold is being used for the first time, you may need to remove the channel frames from the sides to get the plate out. You may have to pry it a bit with putty knife or something like that. After a few melts the mold usually drops the plates very easily. For this reason you may want

Casting Plates

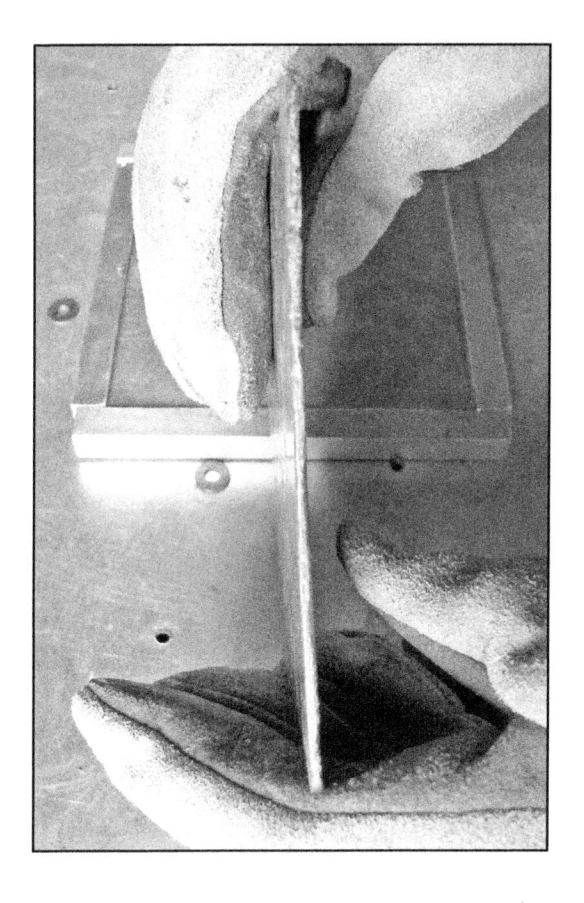

to do a few seasoning pours and then proceed to the melt of your final plates. Practice runs are always a good idea, and will give you a feel for the process.

If you pour from a melting pot into the mold, fill the mold in one continuous pour for the whole plate. Hesitation can create defects serious enough that the plate will have to be melted down and re-poured. This is because the first part of the pour will oxidize slightly, which creates a layer on the surface. The second part of a hesitant pour will not unite with the first pour as well as it should, and this creates a weakness in the plate. Usually this is more of a problem and more noticeable when pouring into a cooler mold; but problems can still occur from inconsistent pouring into a hot mold.

You can get wavelets of dross on the top as you are not skimming the surface to remove oxides and impurities. Do not be concerned about this as the plate when filed lightly will remove the imperfections easily. If you screed the surface most dross will be removed.

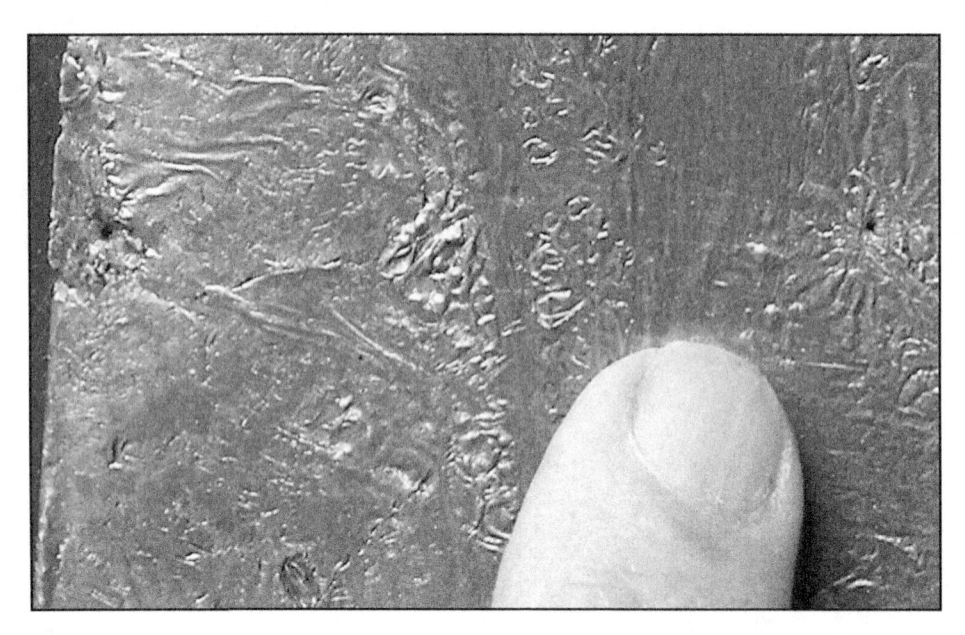

Other imperfections such as small pits can occur but they are not of major concern. You can either remelt or leave well enough alone and live with them.

When melting lead in a pot for a pour, be sure to skim for dross constantly to remove as much oxide as possible before pouring into the mold. A spoon will suffice for skimming. Once the lead is in the mold, skimming the plate surface is not easy. There is a tendency to pull the lead and create voids especially when working with thin melts.

Simple plate mold

Another type of mold can be constructed of strips of aluminum or other metal which are cut to the size of the plate needed, and laid on a griddle. The four pieces are aligned on the griddle to form a mold dam. The surface tension of the lead keeps the lead from running out through the cracks at the edges, so the strips do not have to be fastened to

Casting Plates

each other. For the example in the photos I used 1/16" thick strips that were 1/2" wide to create this easy form. This works very well and does not require as much work to construct as the other form. The griddle top serves as the plate mold surface.

Finishing the casting

If there are major imperfections in a cast plate, they can be fixed by flipping the plate over in the mold and remelting. Before you do this, file off the dross on each side of the plate. This process can be repeated as needed.

There are many ways to repair and finish cast lead. In some cases, hammering with a mallet and a piece of metal stock will help to remove mounds and unevenness, but this is not always the best choice. If you do hammer the cast lead plate, you can remelt it or not. Lead does not work harden as other metals do, but it does change structure internally somewhat and becomes more brittle from hammering. Remelting will bring the structure back to cast metal mode and make a better plate, although it is not necessary.

It is also possible to hammer the metal to enlarge the plate size, but this can leave you with variations in thickness unless done well. The hammered plate is then remelted on a griddle. A mold is not needed as the surface tension of the lead keeps it from melting out of shape. As a matter of fact, any plate you remelt will tend to hold its shape and not drip out of bounds.

The most exacting way to even out plates is to use a small hand operated rolling mill. This is an expensive option, but will produce plates of a uniform, exact thickness.

Casting plates without a mold

Another option for casting plates is to simply pour a sheet, or directly melt the lead on a skillet, to the desired thickness. The casting will conform to the shape of the skillet you are using. Let the sheet harden and then remove it from the griddle. We used a piece of sheet plastic to transfer the cooled lead plate from the griddle to the work surface (see photos next page).

Casting Plates

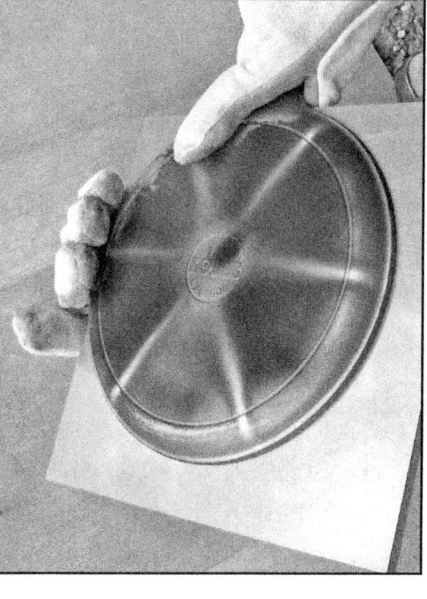

If the plate does not come free easily, try gently prying around the edges, and hammer the back of the griddle.

121

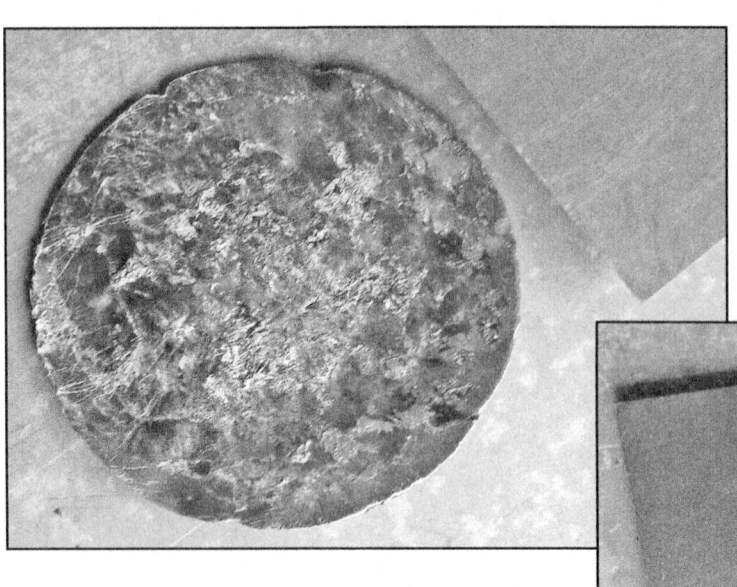

If necessary, the plate can be straightened and flattened with a block and hammer before cutting. The plate can then be cut to size using a template cut from plastic or other material. See page 127 for more information about cutting lead.

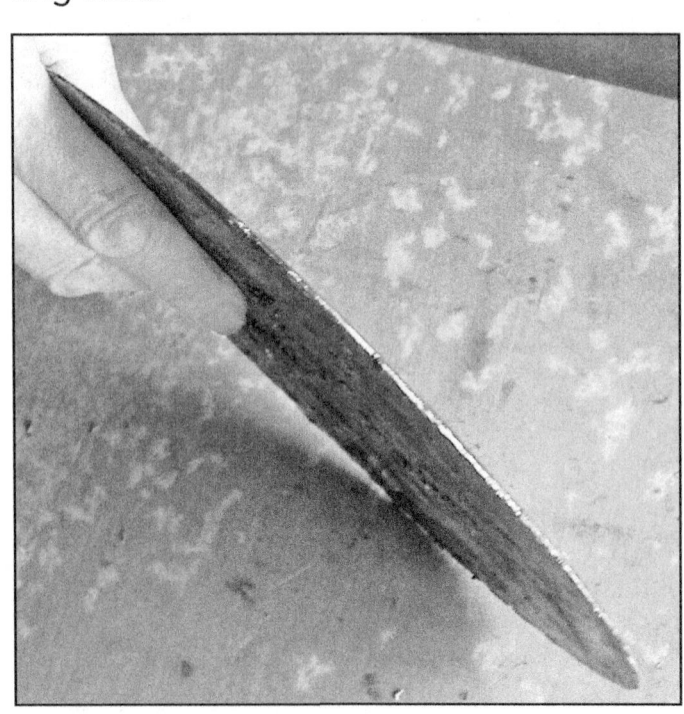

Casting Plates

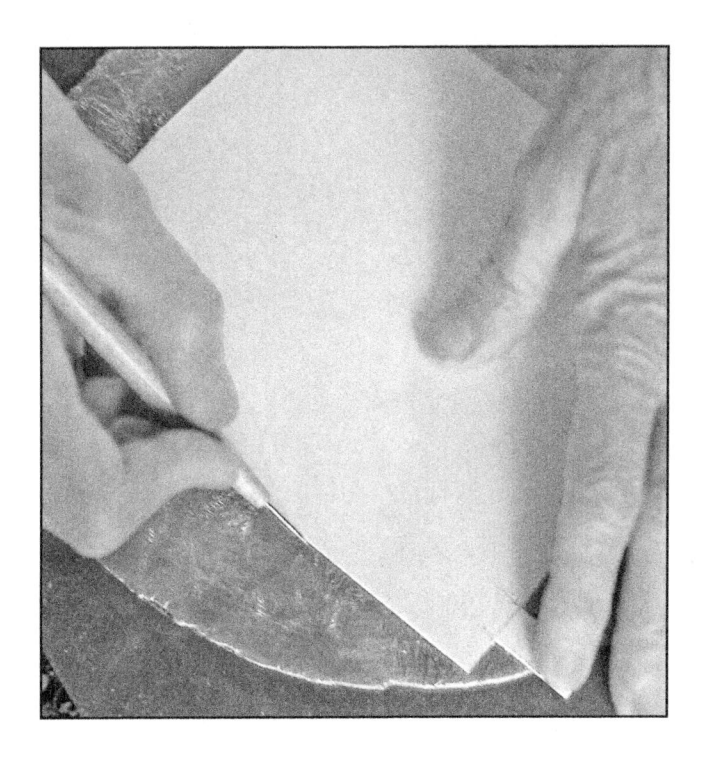

First, the outlined is scribed, then a linoleum knife is used to deepen the scribe lines.

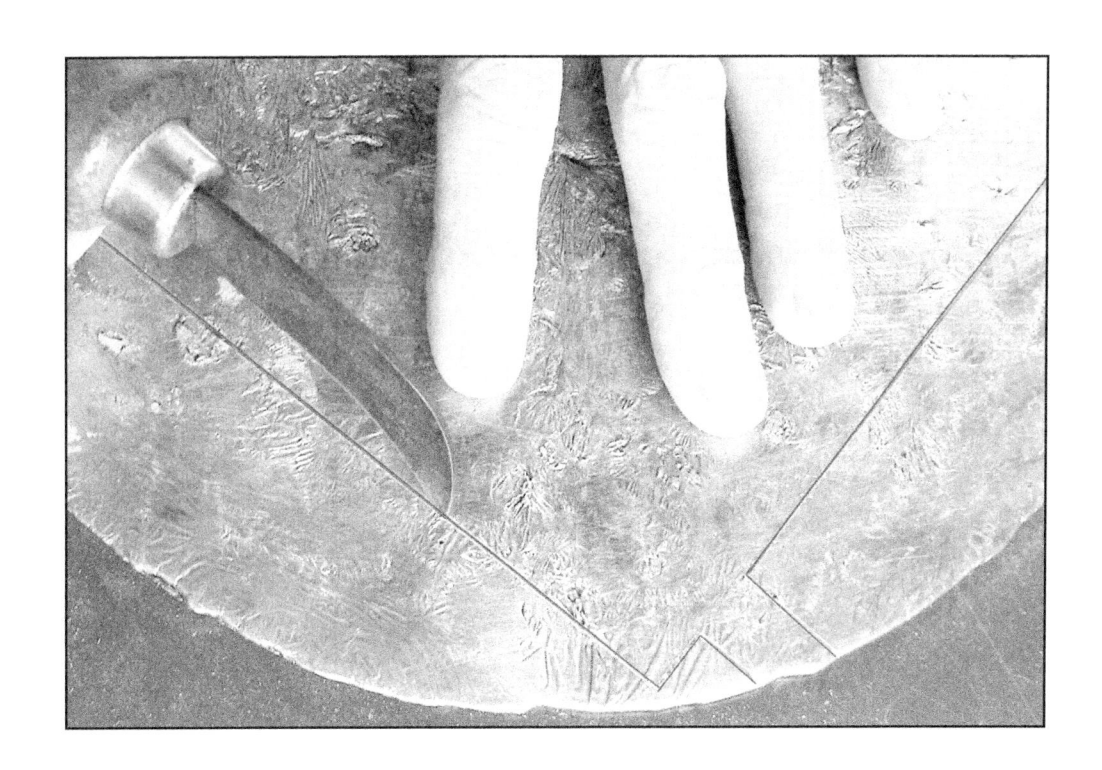

The Battery Builders Guide

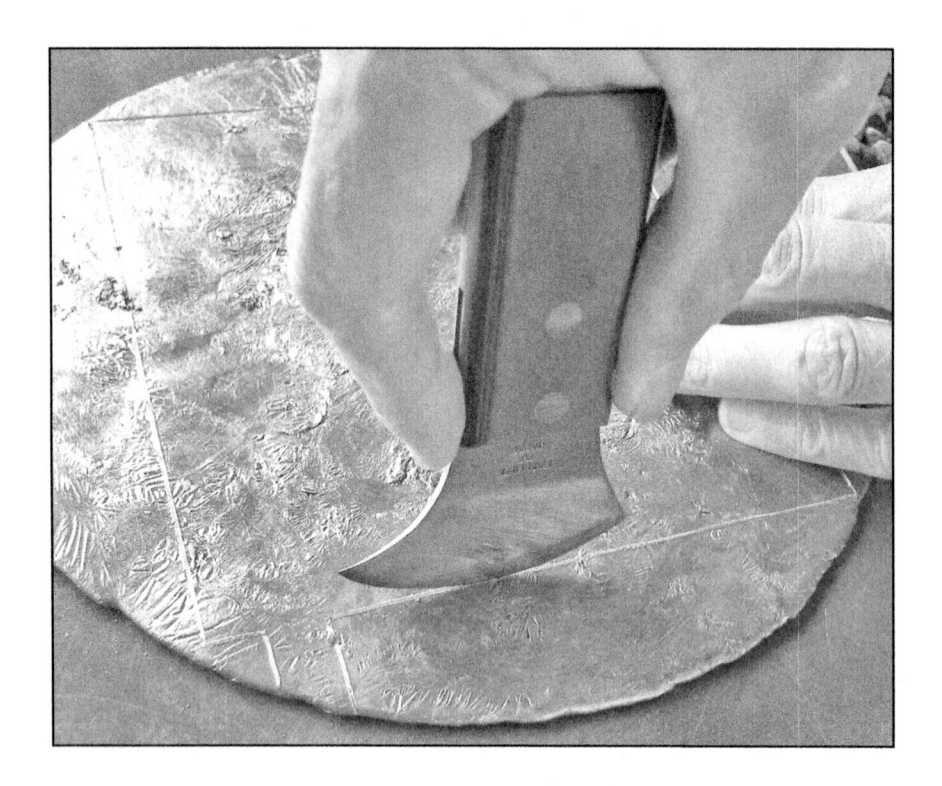

Finally, the lead is cut through with the fanout knife.

Working with Sheet Lead

Tools for Working Sheet Lead

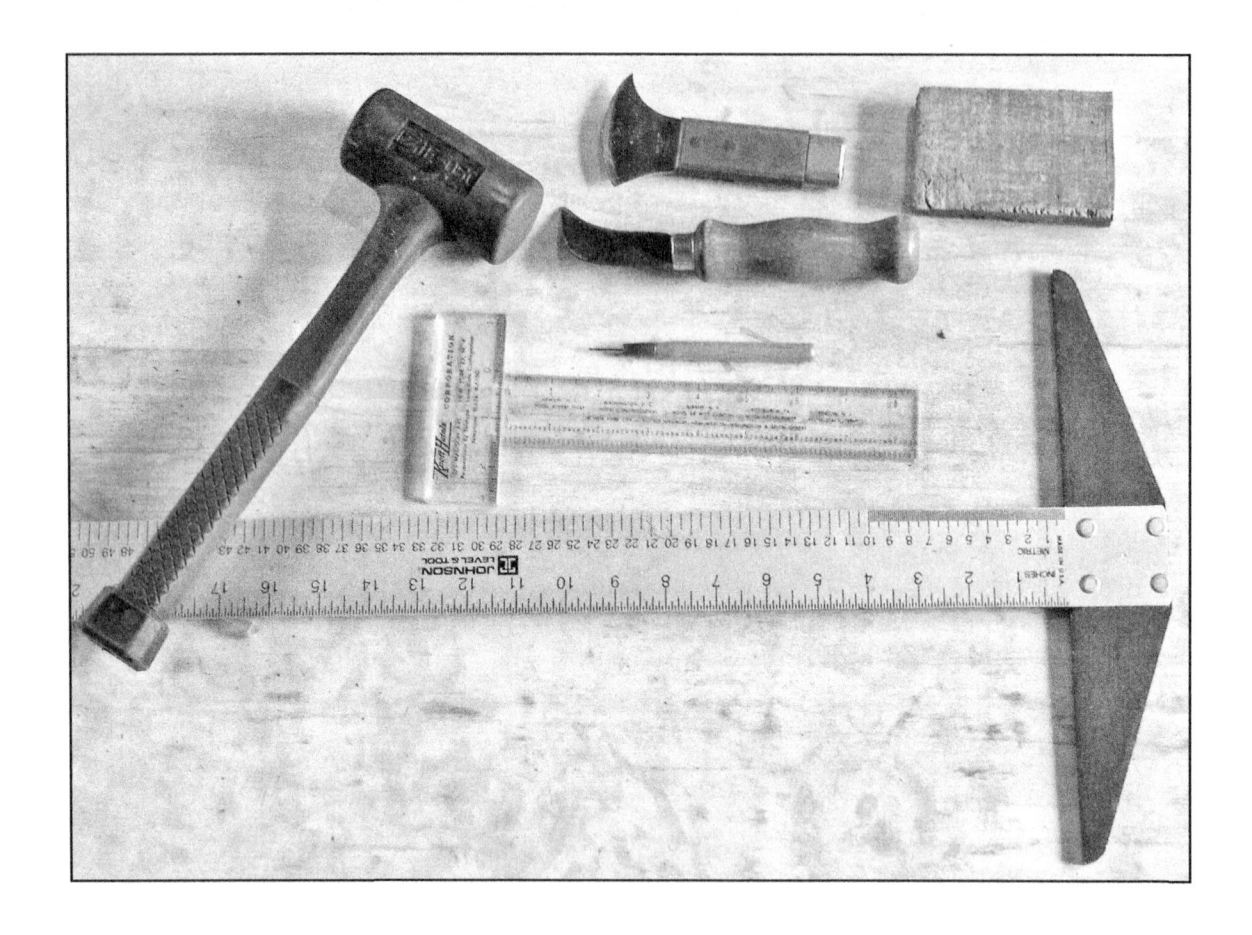

Sheet lead comes in a variety of thicknesses. Thicknesses greater than ⅛" will rarely be used, though on occasion, batteries are made with ¼" plates.

There are four basic functions to perform when working with sheet lead prior to burning: smoothing, measuring, cutting and finishing. For measuring, you should have a variety of rulers, squares, and T squares on hand.

Smoothing sheet lead

A soft faced, lead weighted mallet can be used with a block of wood or PVC pipe to straighten lead sheet. For this purpose, you should have on

Tools for Working Sheet Lead

hand various small blocks of wood, and PVC pipe of varying dimensions.

Lead workers use a variety of wooden shapes for smoothing lead sheet. These devices are called dressers. A wooden darning egg makes an excellent dresser. The egg shape is perfect for smoothing out dents and kinks, but any rounded wooden shape will work well enough also.

The darning egg can be rolled or rubbed over the surface area. It is usually better to roll the egg gently over the lead than it is to rub it, as rubbing can sometimes make more defects than you started with. Much depends on the nature of the kink or dent.

Lead cutting tools

You will need several cutting instruments. A good pair of metal shears will be handy, but the most important cutting tool is a lead knife. The best lead knife for most of the work involved is the fanout knife.

Sheet lead is cut with a fanout knife by using a rocking motion from the palm of the hand. This type of knife cuts through the lead very easily with very little pressure, and stays sharp for a long time. It is also used to cut off casting bleeds from the primary casting and to generally cut surfaces to an even shape. For a cutting surface, if you are working with big rolls of sheet lead or generally need a large smooth work

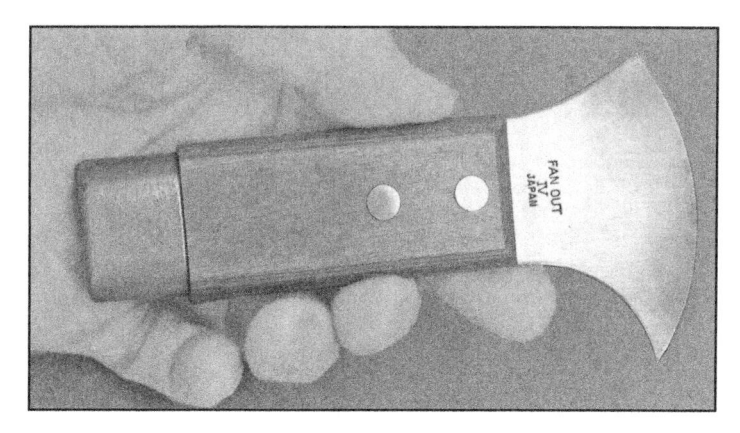

Fanout knife

surface, plywood is probably the easiest and most inexpensive surface to use. It can be laid down on the ground or floor and be used for rolling and smoothing the lead as well as final cutting. Do not use a metal surface for cutting lead as it will dull the knife. Keep a sharpening stone at hand for the knife, and use it.

Making the cuts

To mark a sheet for cutting, first use a metal scribe to score the lead with. Then use a linoleum knife to deepen the scribe line by cutting in along the original scribe line two to three times. The linoleum knife will follow the scribe line quite easily but you want to do this at moderate speed so that you

Mark the cut with a scribe

do not jump up out of the original scribe line. The depth of the scoring will depend on the pressure you put on the linoleum knife. After deepening the scribe line with the linoleum knife I then place the fanout knife within the score line and with a rocking motion from the hand perform the final cut along the line.

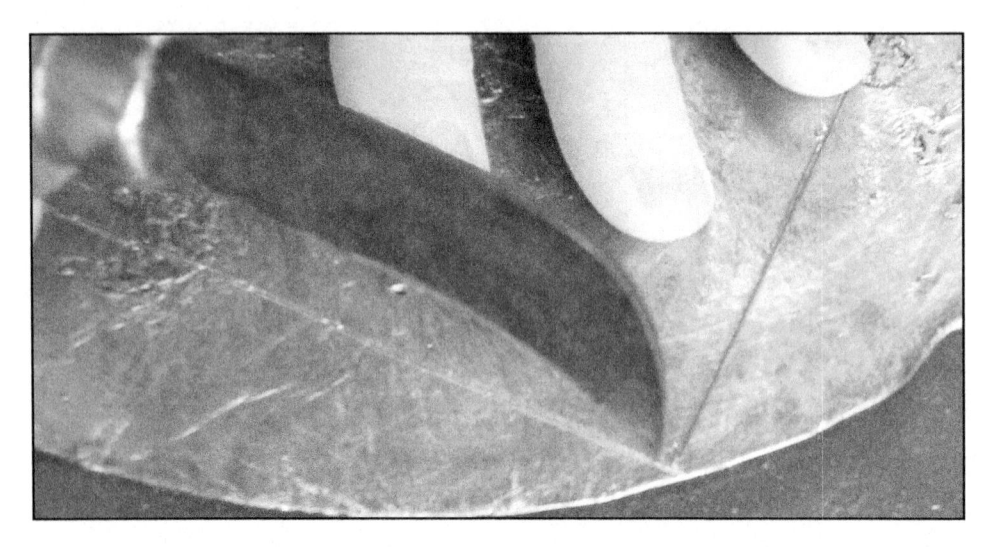

Deepen the score with a linoleum knife

Tools for Working Sheet Lead

Using the fanout knife

The points of the fanout knife are very good for cutting ends and corners precisely. Practice on some scrap pieces and you will master the use of this knife very easily. A word of caution is necessary. This knife is razor sharp. You can seriously cut yourself just by touching the edge of the blade. If you do not pay attention to where your fingers are at all times in the cutting process, you could easily lose part of a finger. In addition to accidentally cutting into your flesh, which would be bad enough, you risk getting lead poisoning from the cut, because lead accumulates on the knife edge while you are cutting. Take this warning seriously. It's wise to wear Kevlar® gloves when working with this knife.

Make a holder for the knife to cover the blade so that you or anyone else can't get cut inadvertently while it is waiting to be used. This is one knife that should never be flung into a box of tools.

Making Plates from Lead Sheet

Smoothing the lead

Whether you use new or recycled lead sheet for making plates, it will need to be straightened and smoothed out. New sheet is not likely to go through the shipping process without getting somewhat mangled, especially at the ends. Always plan to have to work on the sheet to smooth it out. If you purchase recycled sheet you can inspect the sheet before you purchase which will give you a good idea of how much work will be involved to smooth it.

A roll of new lead sheet fresh out of the delivery box.

To smooth out sheet lead you need a large flat surface to work on, such as a large piece of plywood. Lay the sheet on the plywood workspace and gently unroll the piece you are working with making sure not to unroll squished spots with deep dents too fast as this could rip the lead. Work slowly to open the sheet or pieces to full size. I use a long, straight and smooth stick such as a paint roller extension pole to begin the unrolling process. Push the sheet down gently with

Making Plates from Sheet Lead

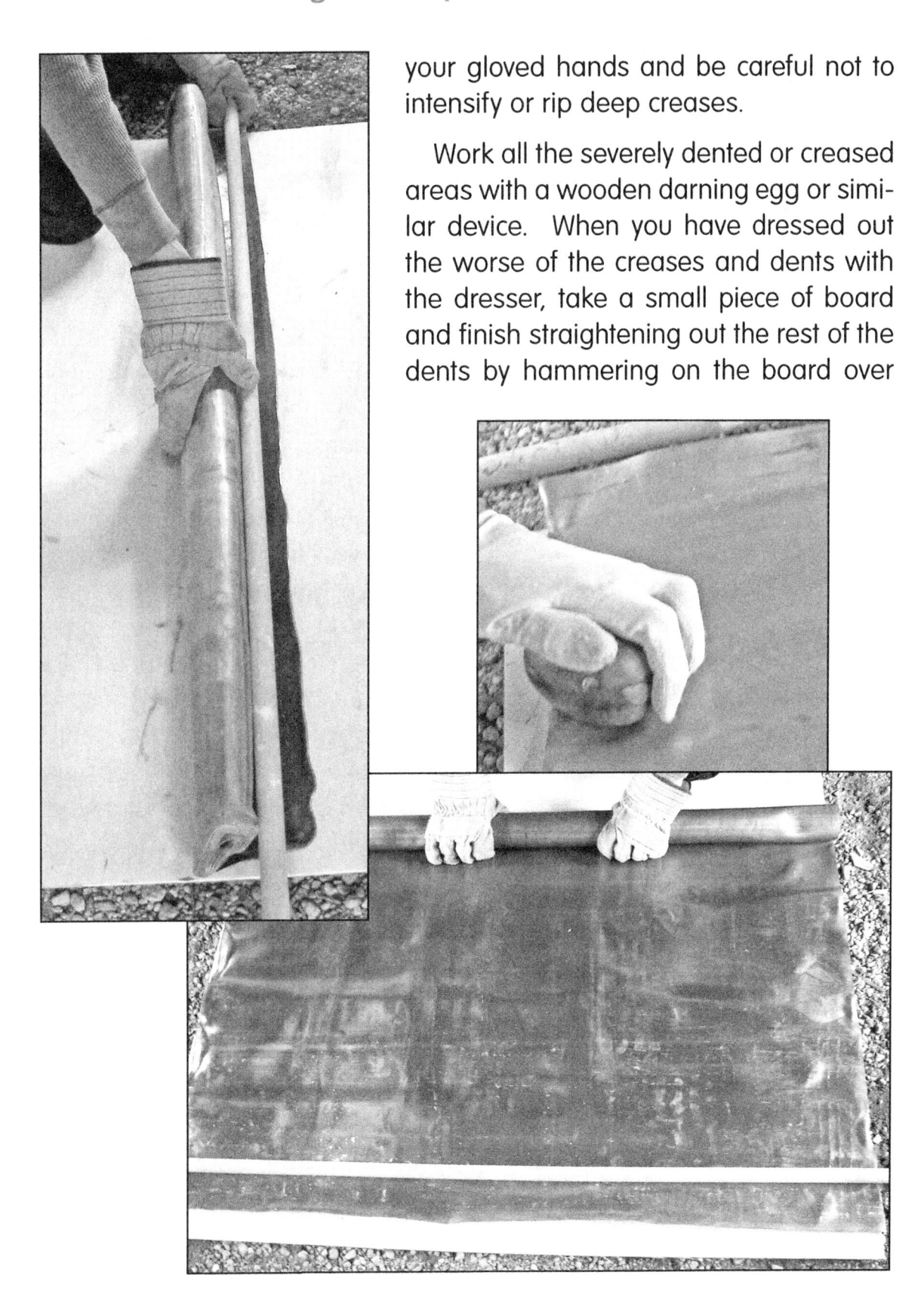

your gloved hands and be careful not to intensify or rip deep creases.

Work all the severely dented or creased areas with a wooden darning egg or similar device. When you have dressed out the worse of the creases and dents with the dresser, take a small piece of board and finish straightening out the rest of the dents by hammering on the board over the crease, gently bringing it back into shape.

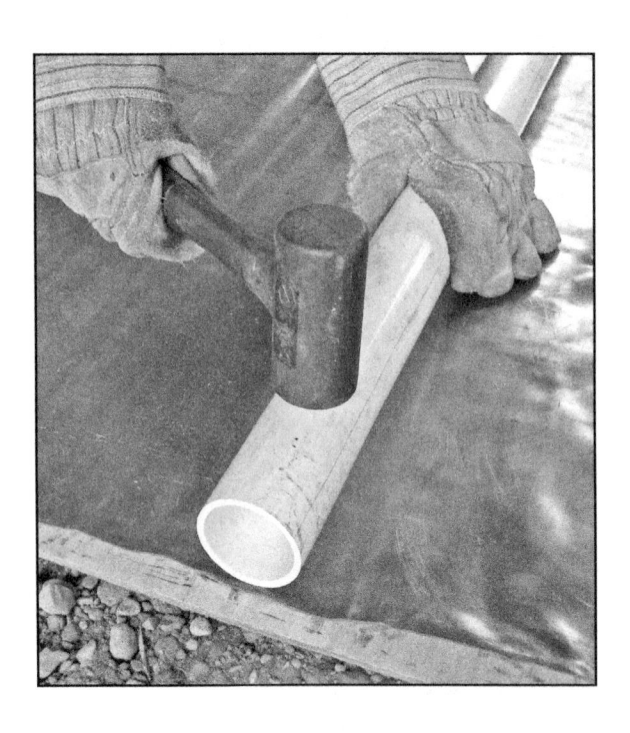

When working with boards and hammering, do not let the edge of the board create a dent in the lead. Remember that all your actions should leave a smoother surface without adding any scratches or dents.

For large sheets, the next step is to take a PVC pipe and roll it over the sheet to even it out. Press down very firmly while doing this. The PVC pipe should be longer than the longest dimension of the sheet lead. If not, the edge of the pipe would dent the lead as it was rolled across the sheet. After the first roll or two I then roll the pipe over the sheet and gently hammer the pipe against the sheet to further the process. At this point I usually go back to the wood block and mallet and gently work the sheet over its entire surface and make sure that the edges are straightened out as much as possible. You will be quite surprised at what you can do with a banged up sheet of lead after working it for a while.

To prepare to layout and cut the plates, start with a square corner of the sheet stock.

Make a template for the plates. If you are using a recycled case, the model for the template can be one of the old negative plates from the recycled battery. This will give you the exact dimensions for a perfect fit. If you don't have an old plate that goes with your case, you will have to base the template on the dimensions of your case. The template can be made from plastic sign material, available at any hardware store. Any easy to cut plastic sign will suffice.

Making Plates from Sheet Lead

If you have a large piece of recycled or purchased sheet lead, you have to plan a layout so that you can get the most out of the sheet.

For example, to build two 6 volt batteries that have seven plates for each cell, (three positives and four negatives), there are three cells in each 6 volt battery, so I will need a total of 42 plates. I have measured the recycled case I am going to use and find that I need plates that are 6.5" wide by 7" plus 1" added for the lug length. Laid out as shown below, I can cut 42 plates from a 48"x 48" sheet of lead.

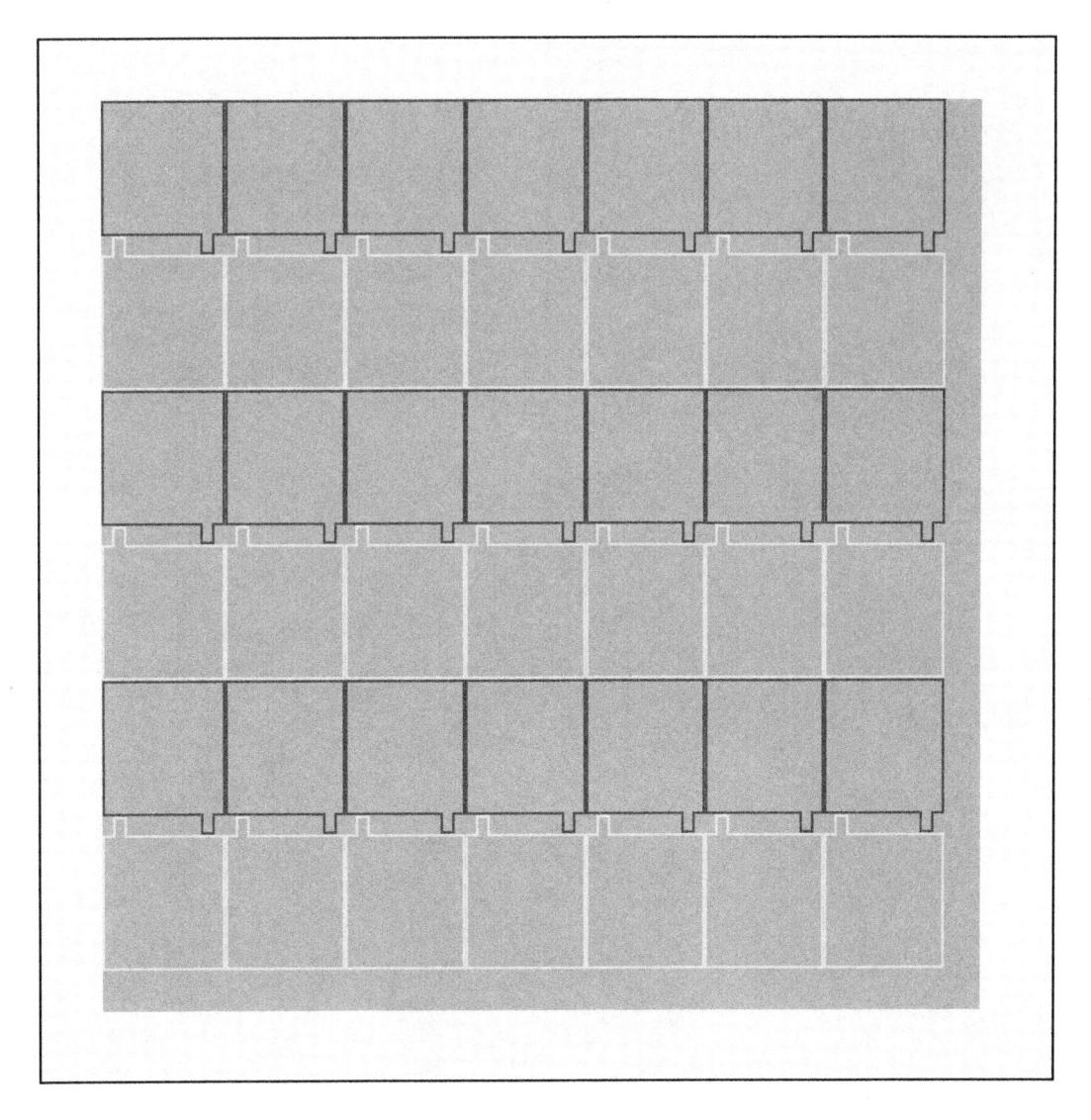

The Battery Builders Guide

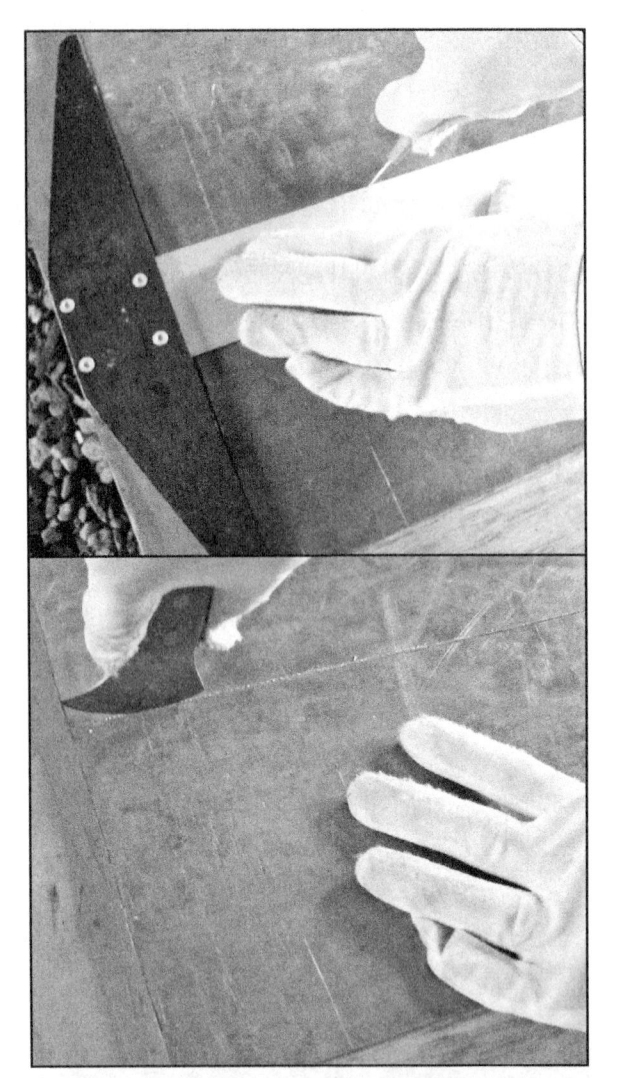

It's easiest to first cut seven long strips that are the width of the plates (the vertical rows in the illustration). With the entire sheet laid out, I started from the squared corner, measured with a T square, marked with a scriber, and cut 7 strips, each 6.5" wide.

The strips were rolled with the PVC pipe to smooth them out. A wood block and hammer can also be used at this point if needed.

Three 15" pieces will be cut from each strip. This will give me 21 pieces. Two plates will be cut from each of these pieces, see illustration next page.

Making Plates from Sheet Lead

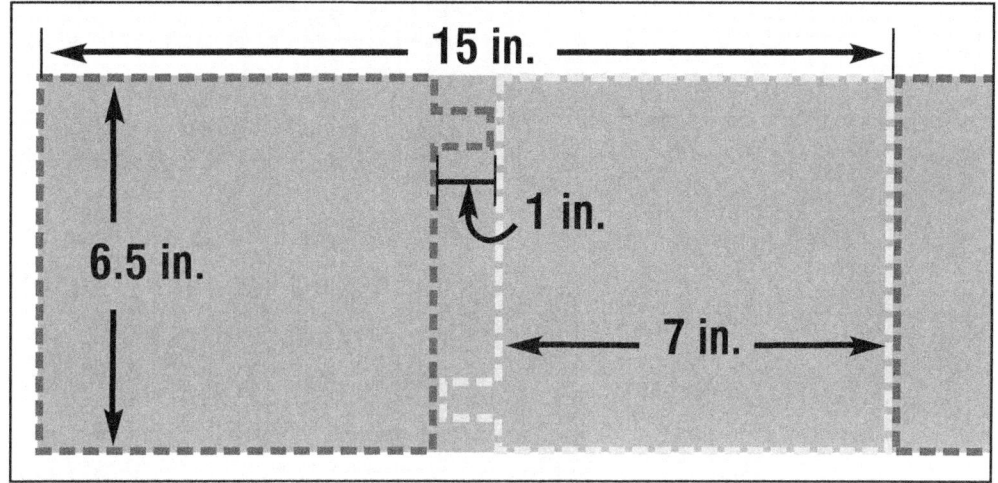

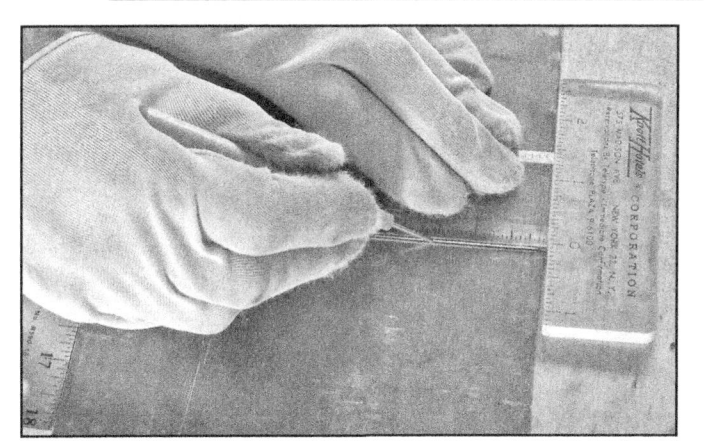

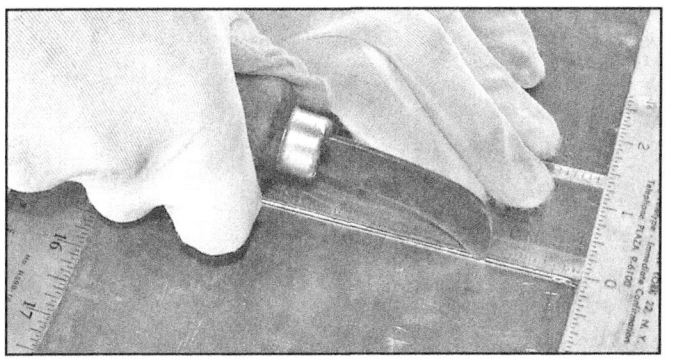

The 15" lengths are measured and marked with the scriber. Since this is a simple straight cut with no corners, the entire cut can be made with the linoleum knife.

135

The piece is scored repeatedly with the linoleum knife, then lifted up and bent back and forth along the score. The two pieces should break cleanly away from each other.

The 15" pieces are smoothed as needed, then the template shape is scribed onto them. When working with templates it is better to use a sharp scribe to mark the plate outline rather than using a pen or other marker. The line from a scribe is more precise, and the groove makes it easier to score and cut.

The outline for two plates is scribed on each piece. The scribe lines are made as deep as possible. The scribe lines are deepened with the linoleum knife, being careful not to stray from the scribe line. The linoleum knife deepens the cuts quite well if you work slowly.

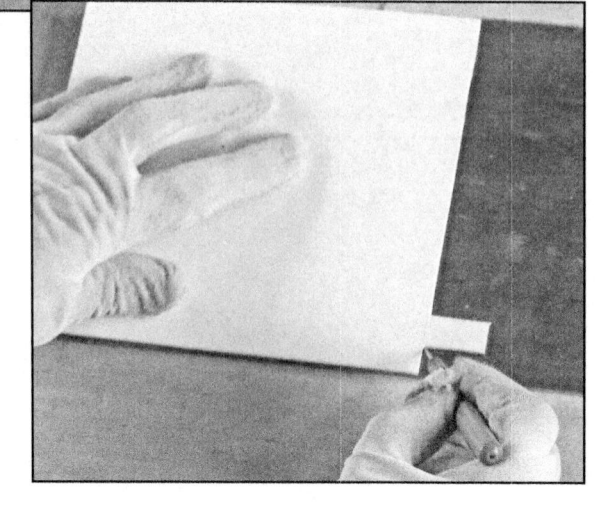

Making Plates from Sheet Lead

The fanout knife is used to finish the cuts.

The Battery Builders Guide

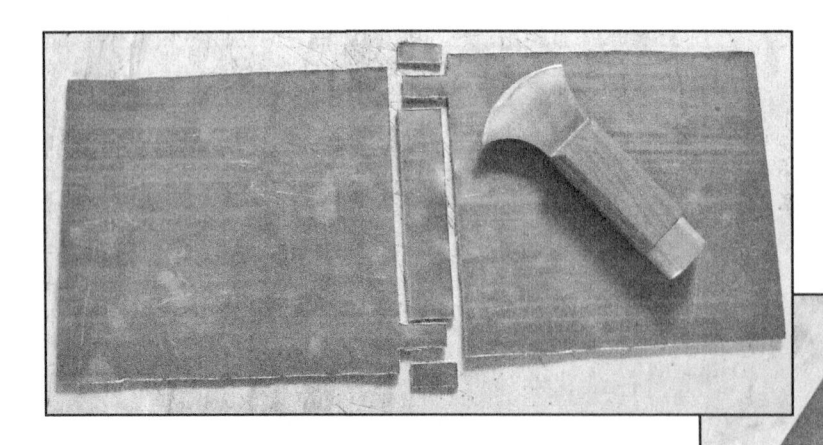

The 42 plates are now ready for finishing.

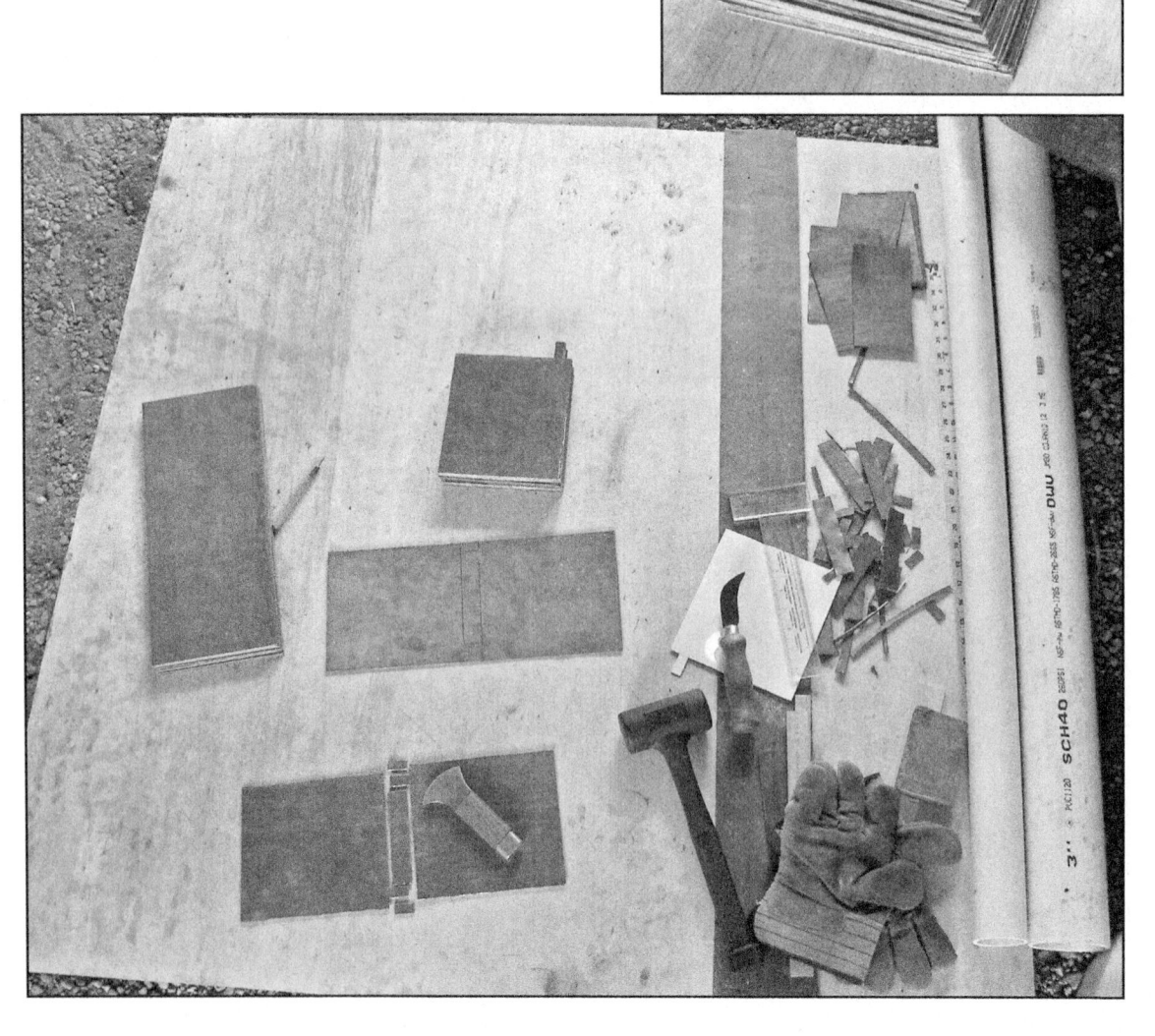

138

Making Plates from Sheet Lead

Finishing the plates

The edges and corners of each plate should be rounded or smoothed by filing to remove burrs and roughness. The plates should be flattened and evened out with a board and mallet. The surfaces can be smoothed with steel wool and a Scotch Brite® type pad for a final finish. Clean off all loose particles and cut off any stringers from the sides of the plate. The plates should be wiped with a damp cloth, or washed and dried.

Adding Plate Texture

After finishing the plate surfaces and edges, the plates can be further processed by either scoring for Planté plates; or punching or drilling grid holes for pasted Fauré plates.

Surface treatment for Planté plates

Solid Planté plates should be scored. The scores in the plate allow the forming oxide to grip the plate better, increase the active surface area of the plate, and help the lead to become porous faster.

The Planté plate formation process takes time, and one thing Planté figured out early on is that if the plates are soaked in a bath of 10% nitric acid for 2 to 6 hours and then washed with a solution 10% sulfuric acid, the forming process would be speeded up. I have not tried this so I cannot comment on this process except to say that you may want to try it if you are so inclined. If you do try this technique, do it before you score the plates.

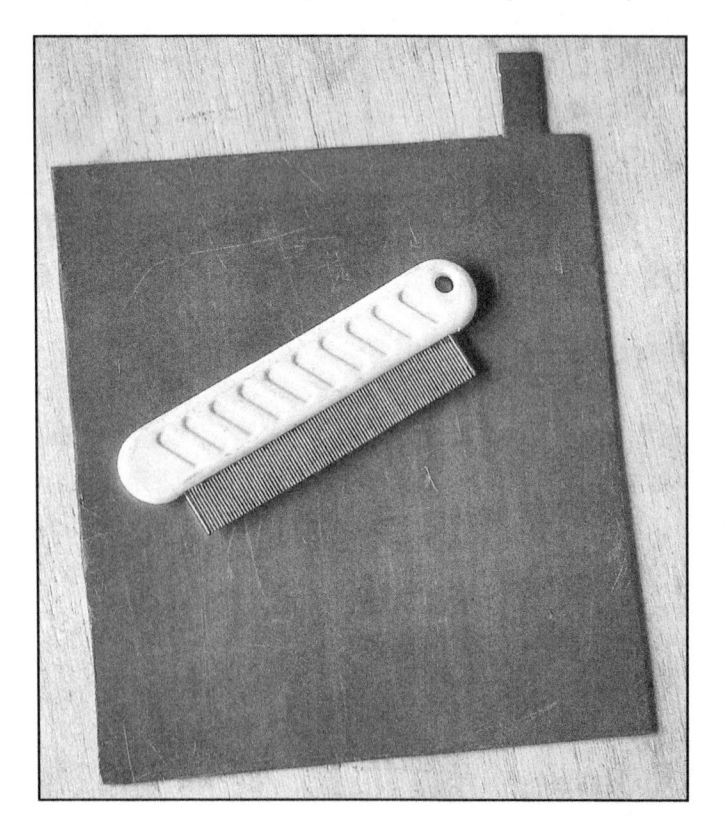

To score the plates, any device that will make grooves up to 1/64" deep will work. It is not necessary to roughen up the lugs. I frequently use a stainless steel flea comb.

If you use a flea comb, press firmly and scratch across the plate from one side to another until the whole plate is scored.

Adding Plate Texture

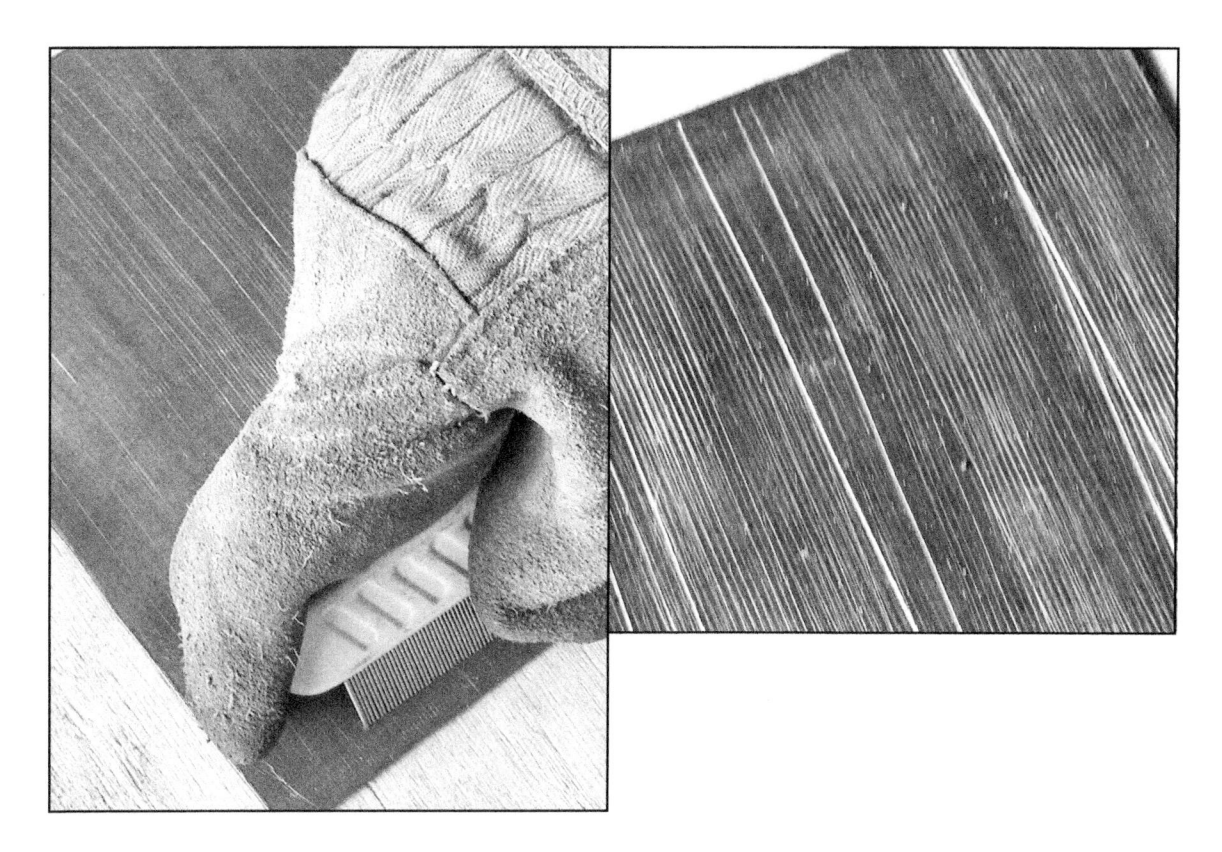

Another technique for roughening up the plate surface is to use a piece of metal screen to impress a pattern into the lead plate. Either aluminum or steel window screen works, but steel is stronger, so it is better.

141

The Battery Builders Guide

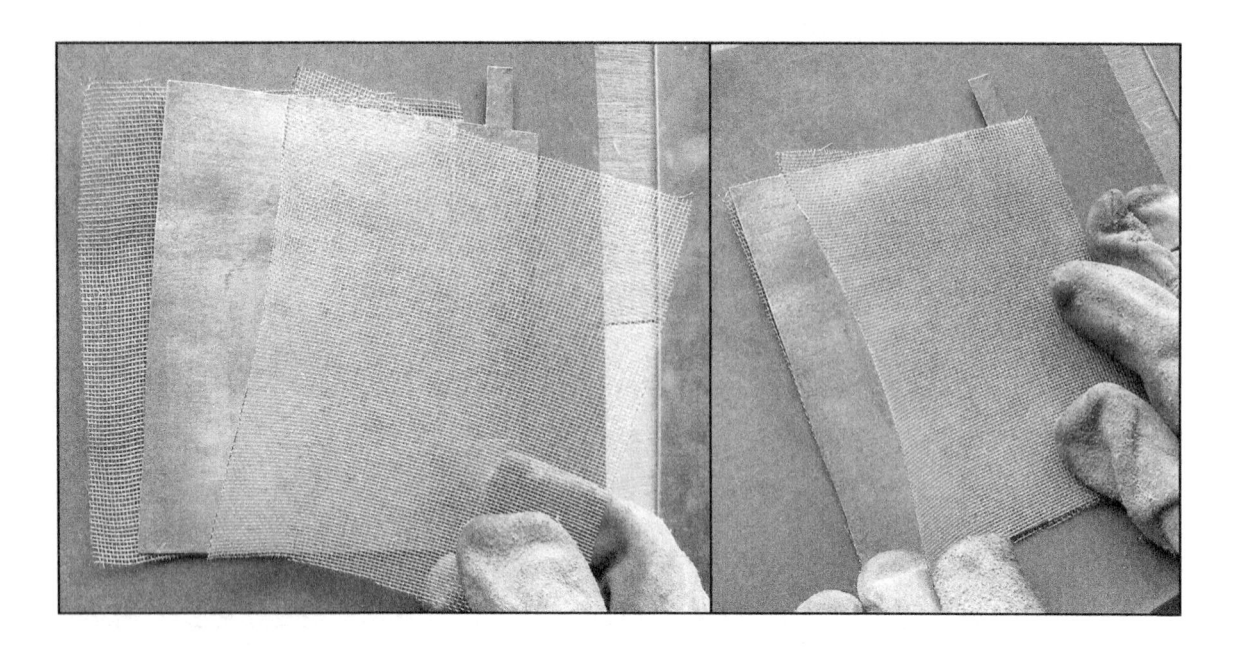

Cut two pieces of screen the size of the plate, and sandwich the plate between the screen on a sturdy plastic or metal sheet. Place a sturdy piece of metal or plastic on top of the sandwich. With a lead mallet, pound evenly across the surface of the plate until the screen is driven into the lead sheet. Turn the plate and screens over and

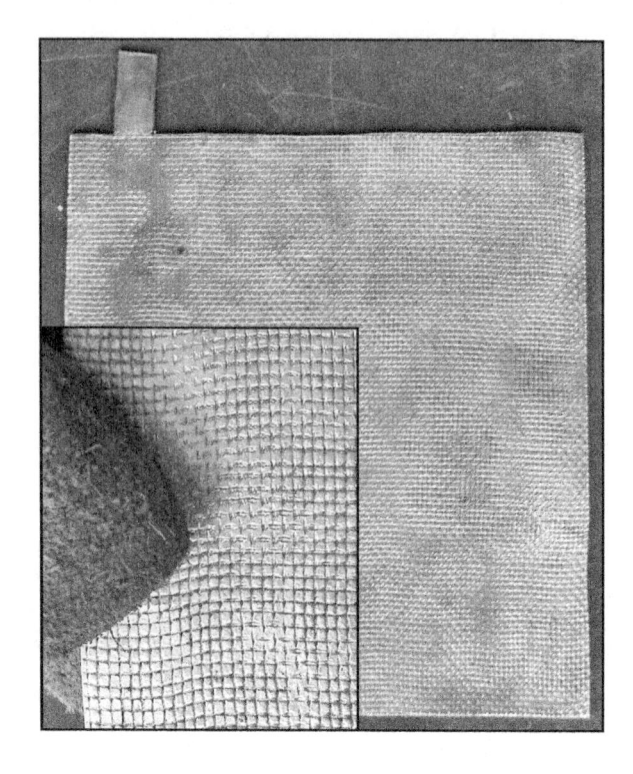

Adding Plate Texture

pound the opposite side. When you are finished the screens will be somewhat imbedded into the lead. They can be peeled off, leaving very nice impressions on both sides. Be sure to clean off all screen particles in case any have broken off. These would contaminate the plates if left on them.

Preparing the Plates for the Burning Rack

At this point, take one of the combs from the plate burning rack and see how the corner of the plate that has to slip into the comb at the bottom, and the lug portion of the plate fit into the slots in the comb. If they do not slide in easily, file the tabs and the relevant bottom corner of the plate until the plate readily slips into the slots of the comb. This saves time when burning the plates to the straps.

It goes much faster to do it all when you're set up for filing lead. After filing, clean the plates and make sure there is no debris left on the plates. The plates are now ready to be burned onto the straps.

Pasted Plates

Making Plate Grids from Sheet Lead

If you are making grids for pasted Fauré plates from sheet lead, after making solid plates as described, you can either punch holes with a hand punch, drill them or stamp them in with a die. The most inexpensive option, but labor intensive, is to stamp each hole. If you drill, you can drill several plates at once with a hand drill, drill press or a milling machine.

Perforated metal or plastic sheet stock can be used as a guide to scribe in the holes for drilling or punching. This can save a lot of time.

Both drilling and punching is labor intensive, so if you are going to make more than a few plates it would be advisable to have a simple die made up that can be used with a mallet or a small hydraulic press to form the holes. Holes should probably be no more than 3/8" diameter if round holes are used. They need to be spaced closely with as little lead in the grid as possible, but still be mindful of the structural integrity of the plates. Of course, the more hole space to hold the lead oxides, the more active area the plates will have.

You will have to figure out the ratio of active area to metal area for your needs (see discussion beginning page 33). If you use unalloyed lead, allow more space between holes because the lead will participate in the electrochemical process and deteriorate. For that reason, unalloyed plates would also do well to be thicker. A lead antimony alloy is best for grids. The alloy does not corrode as readily as pure lead and since it is stronger, the plates can be much thinner.

Mixing and Applying Paste

Pasting plates is a process wherein you make two different pastes, one for the positives and one for the negatives, and apply the paste to the plate grids.

The pasting process is as much art as it is science, and takes a little practice to learn the desirable consistency for pasting. To get a feel for the process it is wise to experiment with the mixing and pasting process on practice grids before attempting to apply your formula to your final selected grids. Also, the quality and properties of oxides vary from supplier to supplier and there can be subtle differences in materials that affect the pasting process.

It is a visual and tactile process and relies on the observation of the texture of the materials; and what this texture means in terms of how it adheres to the grids and the set up time of the paste on the grids. The process is affected by ambient variables such as humidity and temperature, the strength of the electrolyte used in the mix, and the addition of binders and other ingredients. All these factors have an effect on the consistency of the paste, and on the length of time for hardening. The effects of these variables can only be ascertained by you in your particular working conditions.

I generally mix the dry ingredients for one plate, then mix in electrolyte in small increments until the right paste consistency is reached. It is important to thoroughly and evenly mix the dry ingredients before adding the electrolyte.

Using a binder

If one of the dry ingredients is a binder such as CMC (carboxymethylcellulose) you may want to mix the binder with about two to three times its volume of electrolyte before mixing it in with the dry oxides. Mix the binder and electrolyte thoroughly to make a slurry that can be incorporated more easily into the oxides.

CMC needs to be well dispersed and wetted during mixing with electrolyte. The best way to do this is to add the CMC powder in small increments while mixing. Stir the mixture slowly. If you do it the other way around and add electrolyte to the CMC it will tend to produce clumps that are harder to dissolve.

The binder slurry will be poured onto the oxides while mixing. The remaining electrolyte is then added if needed to get the paste to the desired consistency.

Mixing CMC binder into the electrolyte

Mixing in the electrolyte

Adding electrolyte to the dry mix has to be done as fast as is practicable because the mix sets up and hardens quickly.

The amount of electrolyte used is important. If too little electrolyte is used, the mixture will remain crumbly and granular and does not take on a pasty texture and appearance. If this occurs you will have to add more electrolyte in small increments while mixing to the proper consistency.

If too much electrolyte is used a slurry is formed that is too liquid and runny and will not stick to the grids; or if it does stick, it will pull away from the grid supports or crack. The goal is to create a paste that is easy to apply and will adhere to the grids as it dries.

Proper paste consistency, neither granular or runny

Mixing and Applying Paste

When the paste is mixed it should be applied without delay so that the paste sets up and hardens in the grids rather than outside the grids. Thus, it is important to prepare the work area and be set up with all the tools and materials needed for pasting and spreading so that you can work as fast as possible. The only time that pasting would be delayed is if you added too much electrolyte to the paste, in which case you would have to wait for it to set up to the right consistency before application. This remedial delay is best avoided. Always add the electrolyte in small increments and mix each addition in until you get the right consistency.

Paste formula and preparation

There are many paste formulas. Most formulas tolerate a certain range of percentages of each ingredient and still give good results.

The simplest formula is:

Positive plate (cathode)

1 part litharge

4 parts red lead

1 part electrolyte (minimum - add until desired consistency is achieved)

0.2% by weight of total weight of oxide powders of CMC (carboxymethylcellulose) binder (optional)

Negative plate (anode)

6 parts litharge

1 part electrolyte (minimum - add until desired consistency is achieved)

0.2% by weight of total weight of oxide powders of CMC (carboxymethylcellulose) binder (optional)

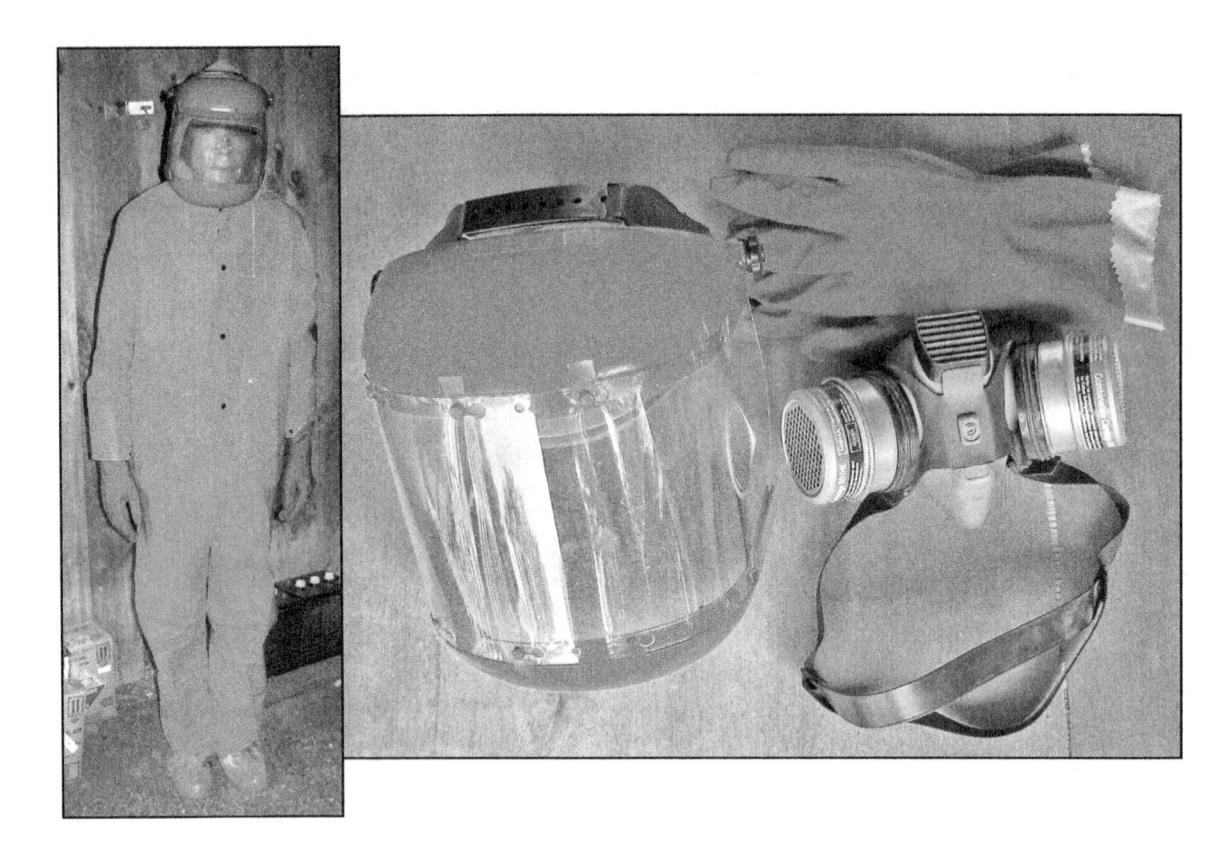

Safety procedures

When mixing paste you need appropriate safety gear. A respirator should always be worn that is rated for working with sulfuric acid vapors and lead oxide dusts. As with all other operations with sulfuric acid you should wear a face shield and total protective clothing, gloves and boots.

Mixing containers and tools need to be of glass, polyethylene, polypropylene, or similar sulfuric acid resistant substances. Do not use metal containers or tools for mixing. A Teflon® or PVC rod is good for mixing.

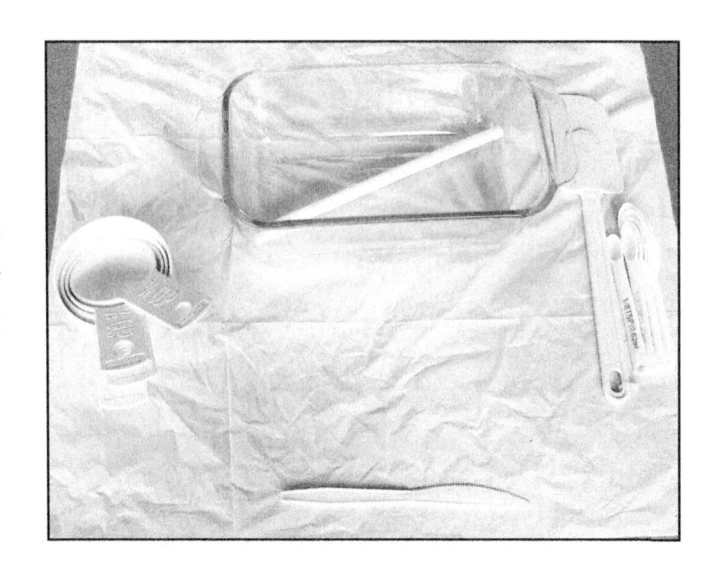

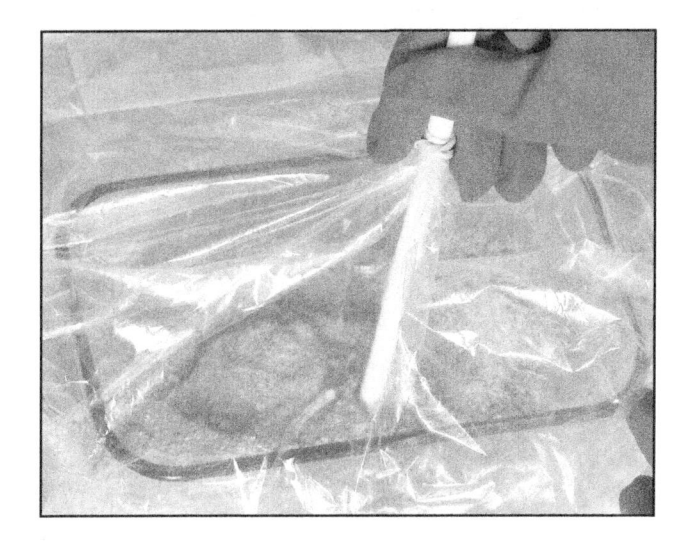

Mixing and Applying Paste

I cover the mixing vessel with a bonnet made from a clear plastic bag with a slit in it to accept the rod, and a rubber band around the rod. This contains the oxide dust during mixing.

A better option is to use a mixing box such as shown below. These are simple to construct and reduce airborne dusts and vapors.

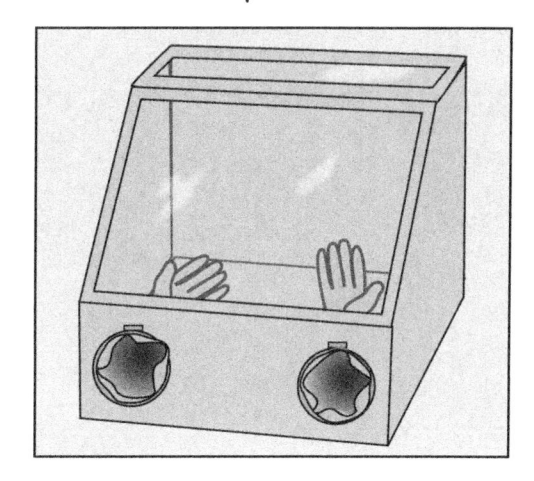

The electrolyte container must be firmly and securely set higher than your mixing platform so that the acid can gravity feed through the dispenser. The platform and container also must be situated so that they do not interfere with other operations.

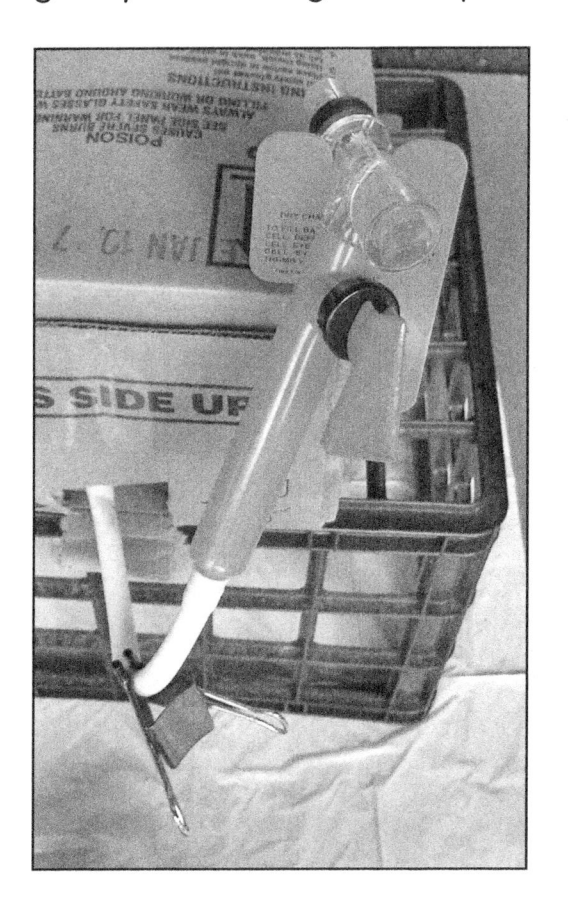

Use a dry charge battery filler with a thumb release valve to control acid flow, or some similar device. Be sure to get acquainted this piece of equipment by practicing with it beforehand. This is a handy device for dispensing acid.

Dry charge battery filler

Set up the work area

Before starting any pasting operation make sure that all the mixing tools, containers, oxides, and electrolyte are laid out so that the pasting process will proceed smoothly. Assess the proper layout for efficient and safe working by practicing dry runs and going through all the motions of the process as if you were actually performing the tasks.

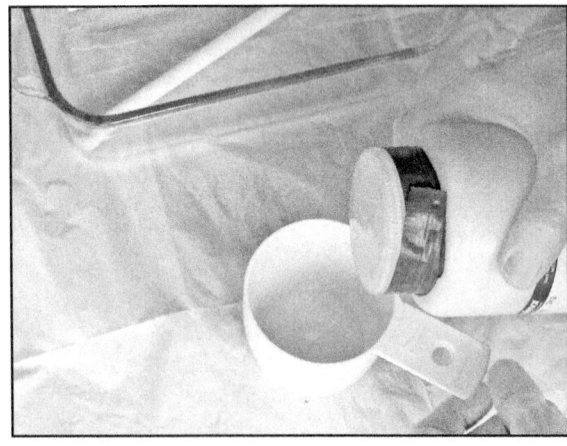

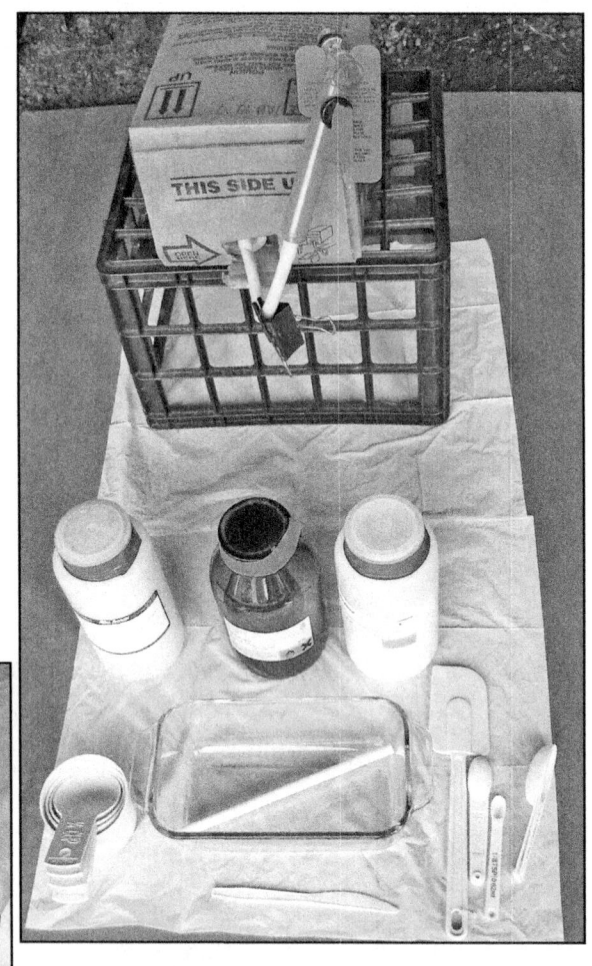

Perform a dry run of the entire process to be sure everything is ready

Pasting container

Before mixing a batch of paste, have the grid plate completely ready for pasting. A kitty litter box makes a good pasting container. It has high enough walls to contain the messy oxides and is acid resistant enough for the short time it is used. A polypropylene, polyethylene or PVC plate can be laid in the bottom of the box to provide a flat surface for the plate if the litter box has mold marks or is curved. The plate pasting surface must be rigid, so sheet should be at least ¼" thick. Do not use any type of metal for this purpose. The pasting plate should be large enough to accommodate the plate being pasted.

Mixing and Applying Paste

Prepare the grid plate

Before laying the grid on the plate pasting surface, be sure the grid is perfectly flat. If not, straighten it by sandwiching it between two rigid pieces of plastic or metal sheets, and hammer the top sheet gently to flatten the grid. You can also use clamps that exert pressure evenly across the surface, like those found in wood shops. Use a stiff and thick material for the sandwich plates so that there will not be any play to interfere with straightening the grid plate.

When the grid plate is perfectly flat, it can be laid on the pasting plate. Plastic wrap can be laid on the pasting plate to protect it so that when you are finished pasting one grid, you will not have to scrape dried paste oxides from the pasting plate before pasting the next grid. If you do this, use a fresh piece of plastic wrap for each plate.

Measuring the ingredients

The quantity of oxides needed for complete plate coverage will vary according to the size and thickness of the plate grid. In general, ¾ to 1½ cups of oxides will be sufficient quantity for most plate sizes. Remember to mix up only enough paste for one plate at a time.

For the first few plates you will have to estimate the quantity of oxides. It is best to overestimate at first and then diminish the quantity after pasting the first plate. For instance, if you estimate that a cup of oxides will cover and fill the plate size you are working with, calculate the quantity of red lead and litharge for the positive plate mix and measure out the proper percentages that you will need within the parameters of 1 cup by volume. Plastic measuring spoon sets and measuring cup sets will work fine for this.

For the negative plate, only litharge is used, so a full cup of litharge is the total oxides needed for that plate.

Above, left and right, measuring the litharge for a negative plate.

Left, any lumps in the dry ingredients must be broken up before adding binder or electrolyte.

When pouring or spooning the oxides from their storage container into a measuring cup, do not tamp down the oxides. When the measuring cup or spoon is full, grade the surface even with the top of the spoon or cup.

Prepare the dry ingredients

After pouring oxides into mixing bowl, break up any lumps and mix thoroughly whether it is a combination of oxides or a single oxide powder. Mixing and stirring will break up the lumps and evenly distribute the powder. When mixing the red lead and litharge make sure to very thoroughly mix these two ingredients together. When it seems like the two oxides are completely blended, mix them more. You really can't over mix, and it is to your benefit to mix more than you think you have to.

Mixing and Applying Paste

Measuring out red lead to add to litharge for a positive plate.

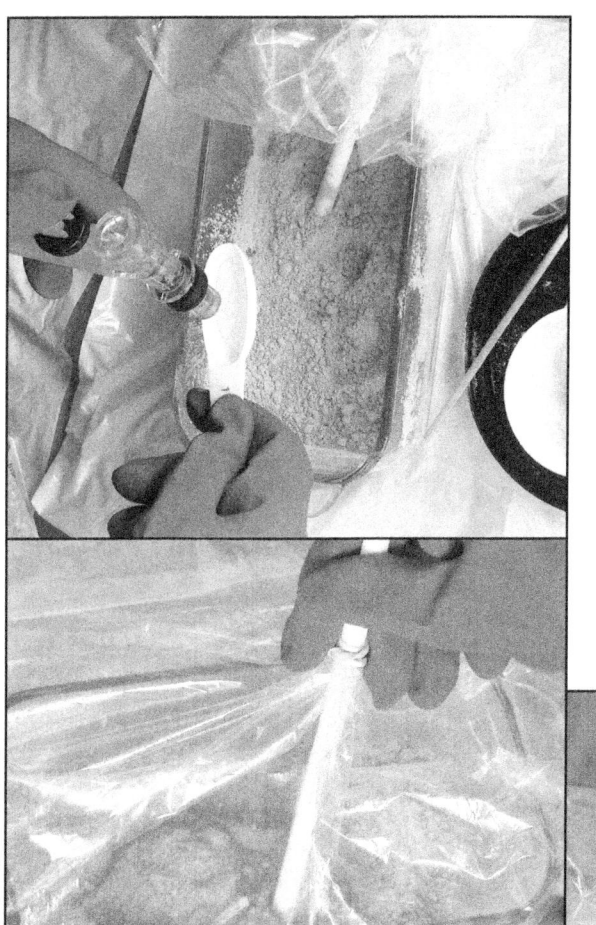

Adding wet ingredients

When you feel that you have thoroughly mixed the dry oxides, the slurry of CMC can be added if a binder will be included in the paste. Pour the slurry in slowly while constantly mixing. When the binder and oxides are completely mixed, the electrolyte can be added in incremental small amounts, mixing in each addition until the paste reaches the right consistency.

155

Mixing the litharge and red lead paste for the positive plate

Mixing the electrolyte into litharge to the proper consistency is a little easier than mixing it into the red lead and litharge. Also, this positive plate mixture of litharge and red lead gives off more heat and undergoes a dramatic color change from bright red to a chocolate brown when the electrolyte is added to it.

Pasting the plates

Spread the paste onto the plate with a spatula. Cover the entire surface of the plate while pressing the paste downward into the grids. It is important for the paste to be firmly pressed into the grid. Use the spatula to remove the gross excess from the top of the plate.

Mixing and Applying Paste

Scrape across the top surface of the plate with a hard, stiff scraper to remove any remaining excess paste from the grids. Flip the plate over and scrape the other side clean as well. Any remaining residue on the grid can be wiped or brushed off, or scraped off with thin sticks or dental tools. Inspect each plate carefully to see that all of the grids are filled and compacted.

Curing and drying the grids

The easiest way to cure and dry pasted grids is to set them aside for about 72 hours where air can reach all surfaces of the grid.

Curing methods vary widely. The grids can be placed in a box or container that is held at high humidity and a temperature of around 86°F for about 72 hours. They are then

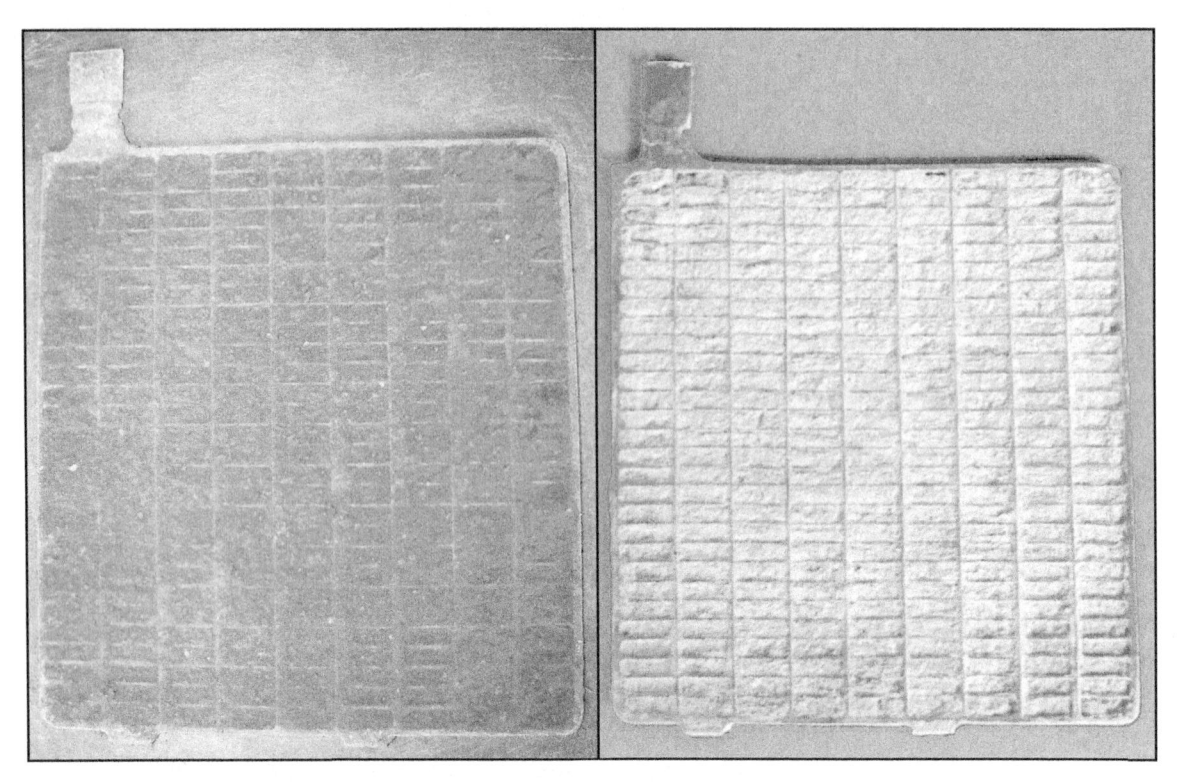

*Dried and cured, the negative pasted plate, left,
and the positive pasted plate, right,*

taken out of the box and air dried for another 72 hours. This technique may or may not improve the quality of the pasted plates. The box could be solar heated with a reservoir of water to humidify and heat the plates.

When the plates are hardened, give the edges and grids a final cleaning if needed. Make sure all dust and loose particles are brushed from the plate.

Store the positive plates separately from the negatives — do not stack negative and positive plates together. Make sure all plates are covered in storage so that they can not be contaminated by debris or dust.

Lead Burning

Plate Burning Rack

Design and function

The purpose of the plate burning rack is to hold the plates of a group at the desired distance from one another, and align the plates so that a plate strap can be positioned and correctly burned onto the plates. Plate burning racks can be designed to accommodate different plate sizes, distances, and thickness.

Our plate rack serves a dual purpose. In addition to holding plates in alignment for burning, it can be used with molds to cast plate slots in poured lead strap connectors (see page 96).

Our plate rack consists of:

- Base and work surface
- Upper guide comb
- Lower guide comb
- Two threaded upright rods with flange nuts to adjust the height of the upper guide bar.
- Two slide rims
- Leveling feet for the base

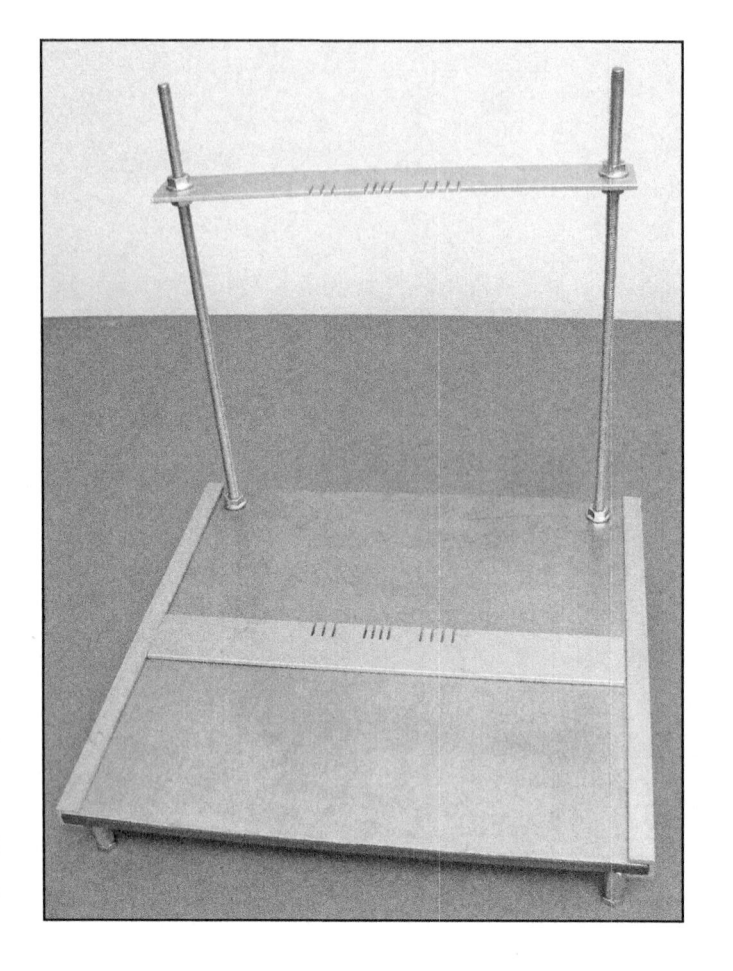

Plate Burning Rack

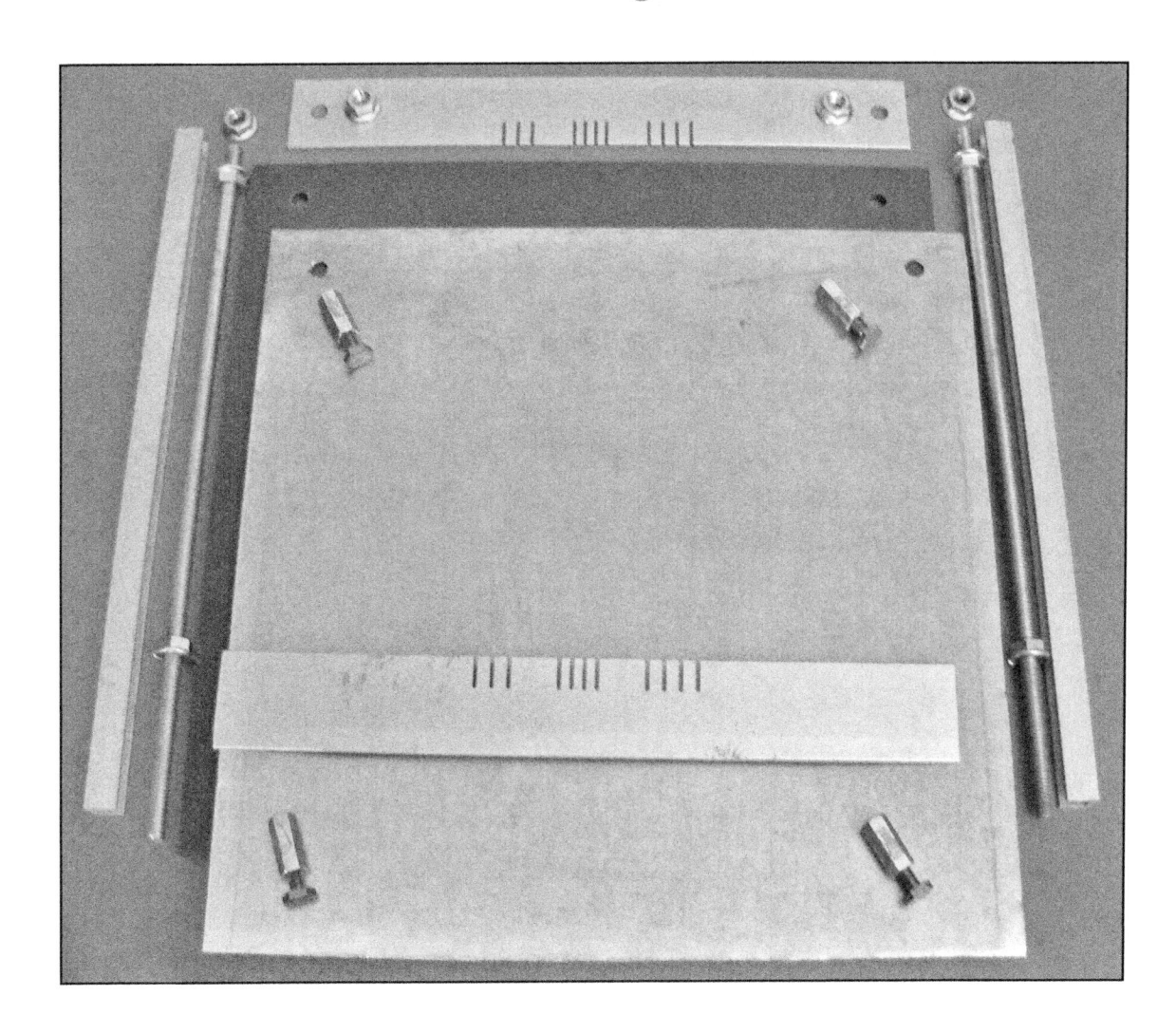

The rack can be built in a variety of ways and with different materials than presented here. Once you understand the principle and purpose, you can easily design a rack to suit your needs using materials of your choosing.

The base of the rack

Our rack was built with a nominal 12"x 12" PVC base. If you buy PVC sheets, be aware that the size may vary from the stated nominal size and you will have to work with the real dimensions of the sheet. The stated width and length tolerance was plus or minus .125". The actual dimensions were 11⅞"x 11⅞". Never assume that the sheet you order will be the exact size indicated. Measure the actual piece before cutting any other components. If you cut the base yourself from a larger sheet, you can cut it to the exact dimensions needed.

For the base, the PVC is covered with a thin piece of sheet metal, cut to the same horizontal dimensions as the PVC. Any sheet metal or aluminum flashing could be used for this, as long as everything fits into the aluminum channel sliders. The idea is to have a very smooth work surface that will not be roughened up by molten lead or errant torch flames.

Although PVC was used for the base of this rack, other materials such as wood, other plastic, or metal could be used. Wood or other plastics would also require a metal cover.

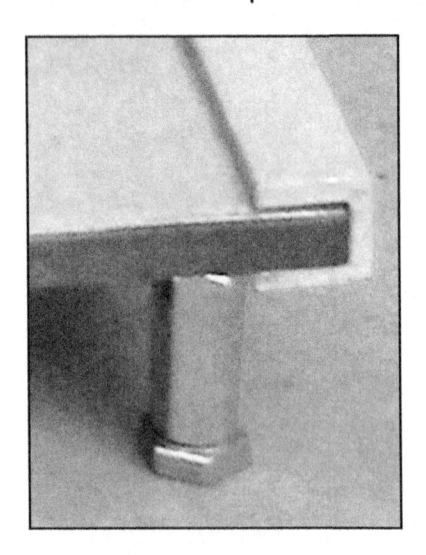

The base and base cover must fit into the slider rims.

Side slider rims

Cut two pieces of U channel to the appropriate length for slide rims and cut shim stock if necessary to tighten the fit of the channel onto the PVC and work surface.

The slide rims can either be epoxied, or drilled and held in place with screws, with shims included, if they are needed. If the fit is extremely tight and it seems like there is a compression fit that will not move out of place, it's not necessary to glue or screw.

Combs

Cut the upper and lower combs to length from a piece of stock 1½" wide and ⅛" thick. The length of the combs should be the exact distance between the edges of the slide rims. For our rack, the length was 11¼₆". The lower comb can slide on the base, tightly guided by the edges of the slide rims to hold different size plates in place. The upper comb rides on the uprights and is cut to the same dimensions as the lower comb so that

Plate Burning Rack

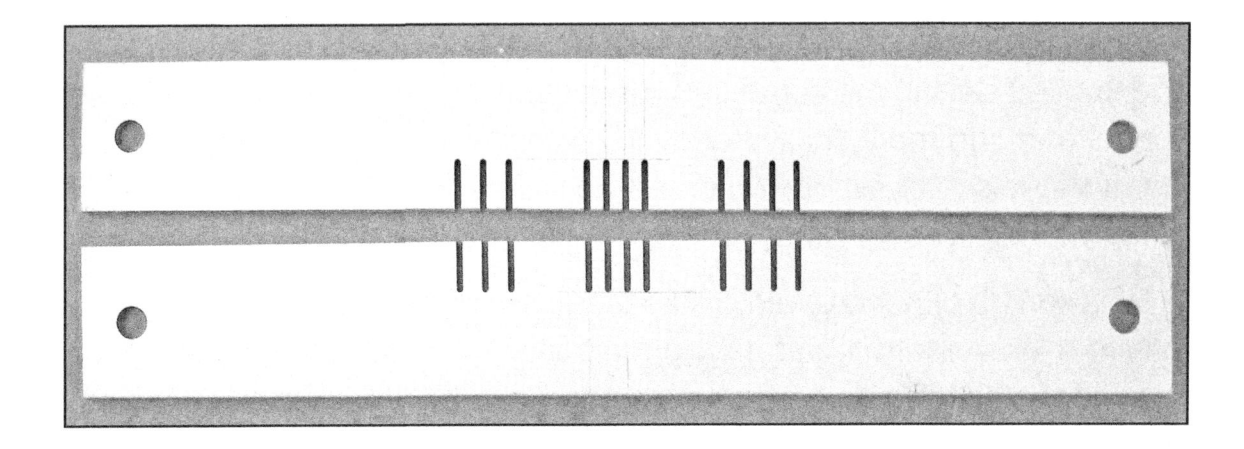

the two combs can easily be aligned exactly. This also makes it easier to match the plate slots of both combs when cutting them.

If the rack will be used for casting slots in poured lead straps, both combs will need holes on each end so that they can be placed on the upright rods. If the rack will be used only for lead burning, the lower comb does not need these holes, since it needs only to sit on the base between the side rims.

Take one of the pieces of bar stock for the combs and mark a drill center point on each end, ½" from the sides and ¾" inch from top and bottom in the center of the 1½" stock. Place the marked comb on the edge of the PVC plate and cover, between the slide rims. If you need holes in the lower comb, place the stock for the lower comb between the upper comb and the base. Clamp all the layers in place and drill a 5/16" hole on each end at your marked points. You could drill the PVC plate, cover and the combs separately but this method avoids errors in alignment of the holes.

Cut slots in the upper and lower comb. The exact number of slots depends on how many plates you will have in the group you have designed. The thickness of the slots will depend on the thickness of the lead plate that you will be using. The distance of these slots from each other will depend on the thickness of the separator material that you choose for your design.

A universal rack can be built by cutting several groups of different sized and spaced slots in the comb sets, or you can cut the slots for a particular project and add more slots later as needed. If you want the option of adding more slots later, when you place the first group of slots, leave enough room on the comb for other groups.

Slots can be either cut with a saw blade or milled in. We milled our slots in with a 1/16" router bit and milling machine. Each comb can either be cut or milled in separately, or the two combs can be clamped together and cut or milled at the same time. Be exacting with the slot placements. The upper bar slots and lower bar slots must align to hold plates properly.

Leveling feet

Glue the metal stand-offs to the bottom of the base, and insert the leveling screws. I substituted hex head screws for the screws that came with the threaded stand-offs to make the leveling feet wider.

Plate Burning Rack

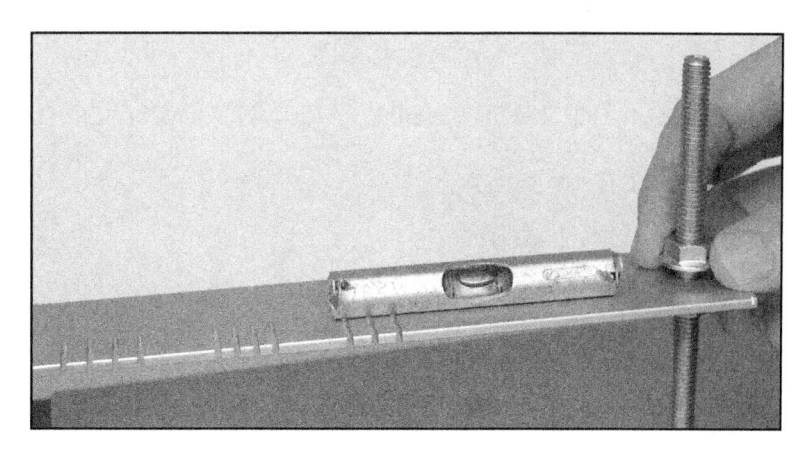

Uprights

For the two uprights, cut a piece of 24" threaded rod in half (12" each). Deburr the edges of the rods so that the nuts can screw on and off easily.

Insert the upright threaded rods through the holes in the base and attach the rods to the base with flange nuts on the top and bottom of plate.

Screw on one flange nut on each threaded rod to support upper comb, and place the upper comb on them. Use a level on the comb and adjust the supporting screws until the comb is level.

Screw on the securing flange nuts to hold the comb in place.

Place lower comb between the slide rims.

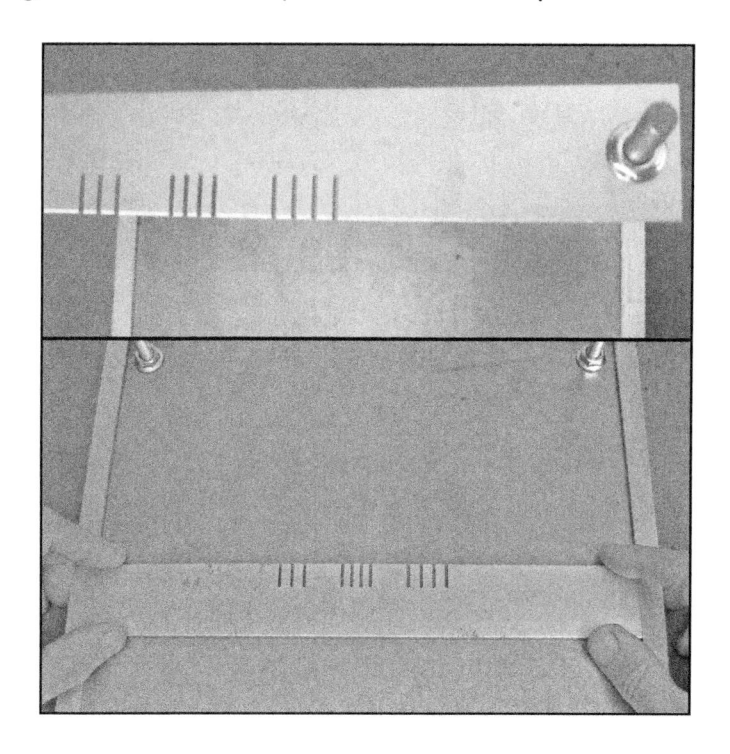

Tools

Available from local hardware store, Micromark or McMaster-Carr:

One 5/16" drill bit

Drill press or hand drill

Saw — hacksaw or other

Router bits — 1/16" diameter, MicroMark #80246. If you need slots for a different plate size, you will need to purchase end mills from McMaster-Carr or MicroMark

Measuring instruments — MicroMark

Materials

Available from local hardware store, or McMaster-Carr unless otherwise noted. McMaster-Carr (MC) item numbers are listed where available:

One 5/16-18 x 24" threaded rod

Eight 5/16-18 flange nuts

One 1½"x ⅛" aluminum bar for upper and lower combs. Length according to your design, about 2'

One 12"x 12" PVC sheet, either ¼", ⅜", or ½", whichever suits your needs. We used ¼" thickness for this particular rack. MC#: ¼" 8747K114, ⅜" 8747K115, ½" 8747K116

One 12"x 12" piece of aluminum flashing, or sheet metal

Aluminum channel 3/32" thick, leg length ½", base length ½" outside dimensions. MC# 9001K31. This material is to be used as slide rims, the dimensions should fit the thickness of whatever base material you use.

Shim stock for aluminum channel for firm fit of channel to plastic base can be any material.

Four wide diameter threaded stand-offs and hex screws to fit them. (Jameco Electronics)

Tools for Lead Burning

Torches

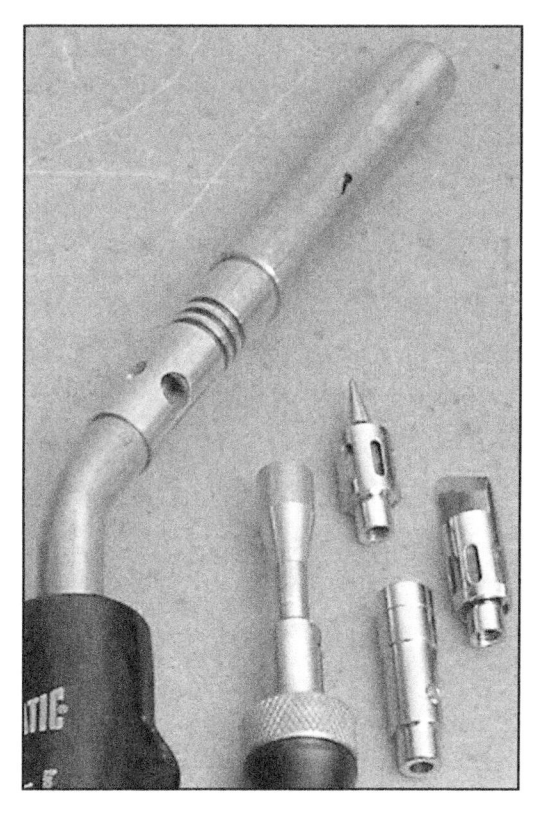

For general plate burning you can use a Bernzomatic® JTH-7 Hand Torch fueled by a regular 14.1 oz propane tank. For lug attachment and other detailed work, use a Bernzomatic® ST900D Mini Torch that also fits the 14.1 oz propane tank.

Both torches have flexible hoses, which are necessary for this work. Do not use a regular torch head attachment that directly connects with the tank. The directly connected torch head will not deliver the gas appropriately at some of the angles necessary for this work. When you use the torch, keep the tank in an upright position at all times. The tanks are light, and sometimes pulling on the hose will tip them, so the tank should be secured into position to avoid this annoyance while working. You will also need an igniter on hand to light the stove and torches.

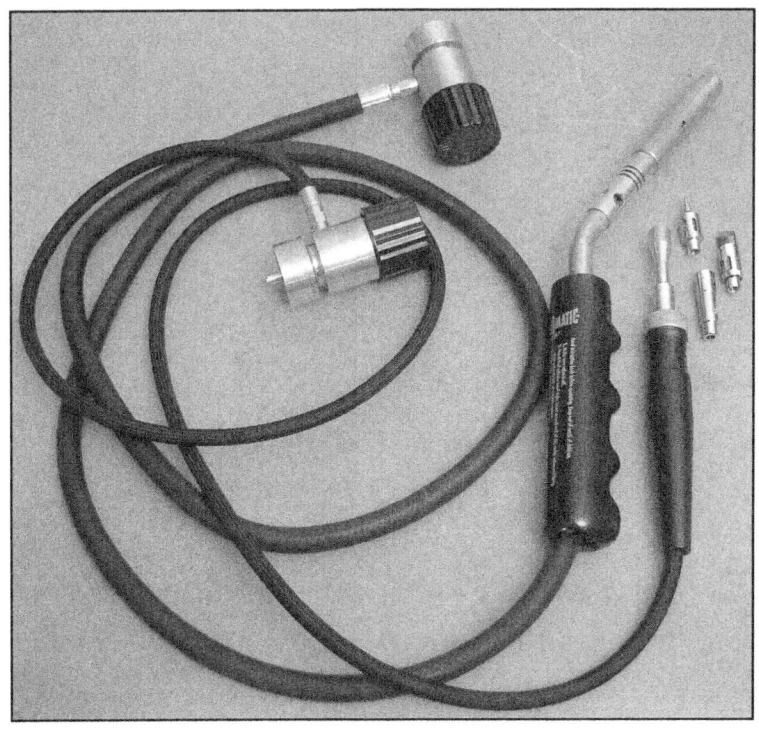

Putty for lead burning

You will also need casting retainer or babbitt putty. The melting point of this putty is about 978°F which makes it ideal for some lead casting molds and for making dams for plate burning.

This putty is very conformal for irregular castings and is easy to use for plate burning dams. The heat of the torch will burn the putty a bit and you will have to blow the flames out from time to time. Try to keep the torch flame away from the putty as much as possible, but you can't really avoid this altogether.

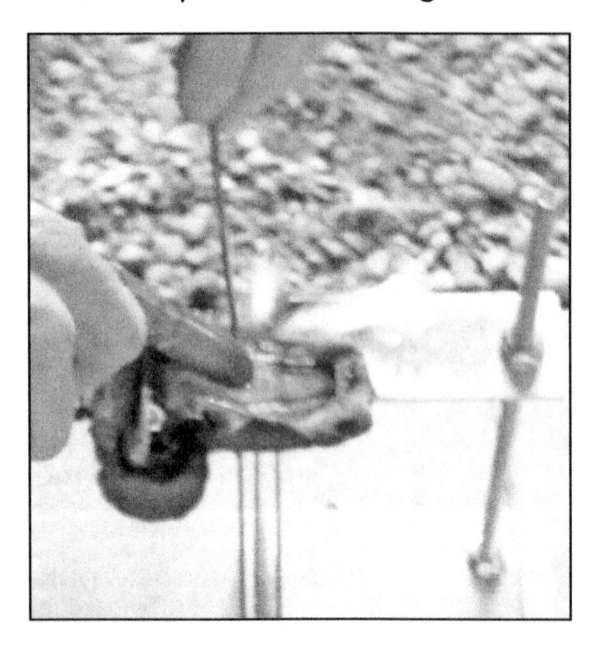

Babbitt or casting retainer putty

When you have finished a burn and take the putty off the work piece, if you remove the part of the putty that's burned from the rest, the remainder can be re-used for the next burn.

Preparing surfaces for burning

Before joining any parts by burning, scrape the surfaces to be joined to remove all the oxides. This allows the metals to join easily upon melting. Scrape thoroughly, down to clean metal.

The tools you use to scrape will depend on the piece and its contours. You can use steel brushes, a triangular carpenter's scraper, files, steel wool, Teflon® scrubbers, or anything else that will get the surface down to bare metal. Keep an assortment of these near at hand for a variety of foundry tasks, including one very large file and a set of smaller files. The smaller files fit into the slots of strap castings to clean them prior to burn-

Tools for Lead Burning

ing. For Planté plates, a very large file can also scrape clean an entire plate before laying in striations or impressing the final groove pattern. Always wear a respirator when abrading lead metal. You can also use a glove box so that any lead dust from your operations is easily contained.

Cleaning the work pieces

After abrading a work piece in any manner, clean it to remove any fine particles clinging to the surface. If you use steel wool or sandpaper make sure that your parts are cleaned well. Use a wet cloth; or you can immerse the work piece in water, wipe it dry and then wipe it down again with a wet or damp cloth. Wet cotton swabs are useful to get into tight places. All parts that will go into a battery must be cleaned well to avoid contaminating the electrolyte.

I find that a good scraping and cleaning is all that you need when working with lead. The only time I use flux is to use ruby fluid on strap tab pieces that are being joined, or when attaching a lug to a plate with the torch. I apply the liquid ruby flux sparingly with a small artists brush.

You should have some heat and metal splash resistant cloth on hand — usually a vermiculite coated fiberglass or carbon fiber fabric. Small pieces are cut to provide heat shields for posts and edges on straps while burning to prevent lead from melting in unwanted places and ruining the work piece. The fabric can also be used for fire resistant clothing overlays to protect yourself and anything else in the area that could be damaged by an errant torch flame.

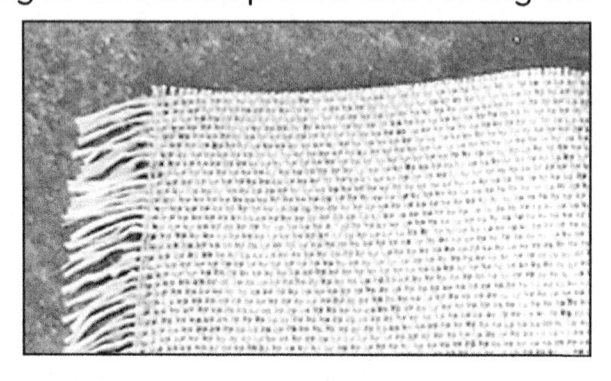

Vermiculite coated fiberglass fabric

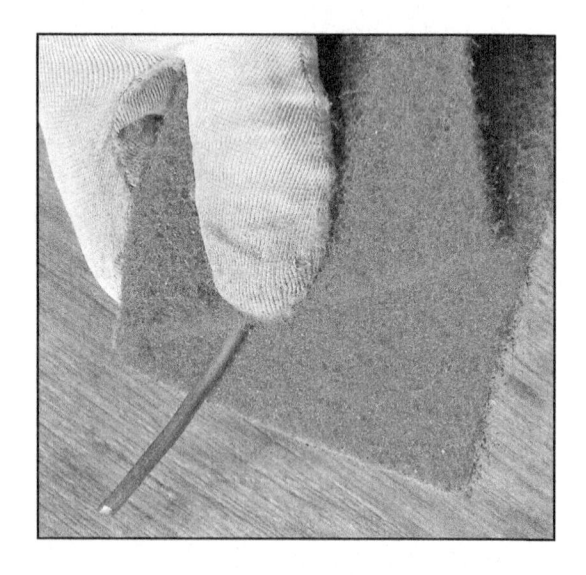

Lead wire

Another important item is lead wire which is used for welding/burning, and filling. Before using lead wire or burning rod make sure to clean off the oxides with a Teflon® scrubber or similar device.

Post molds

Battery post molds are useful lead burning tools found in most battery shops. They are available in either aluminum or graphite impregnated steel and are used to reform battery posts that have broken off during use. The post molds are laid over the cleaned and scraped broken battery post. The top of the broken battery post is heated with a torch until the surface melts. Then lead wire is inserted and melted into the mold until the melt reaches the top of the mold. Please note that these molds should only be used when a post and its group has been removed from the battery well. Never attempt to repair a post on a battery without first removing the group from the case. The hydrogen and oxygen emitted from the battery coupled with a torch flame could very easily cause an explosion. I have seen posts repaired on the battery after wet rags were placed over the battery to act as flame arrestors, but this practice is too dangerous and is not worth the risk of a battery explosion with flying sulfuric acid.

Battery post molds

Burning Lugs onto Plates

If you have cut plates without the lugs from sheet lead, or you are recycling grid plates, you will need to burn lugs onto the plates.

There are basically two ways to do this. Lugs can be cut from lead sheet stock of the appropriate thickness and size, and burned onto the plate or the existing lug remnant; or a dam can be set up in the shape of the lug and a melt laid in.

Burning lugs

If you are burning lugs onto recycled grid plates, always use a lug piece that is longer than you will need. This is because when the lugs were cut from the original strap, they were most likely not cut at precisely the same point, so there will be variations in length. After the lugs have been added to all the grid plates for your project, you can cut all the lugs to the

Burning Lugs onto Plates

Adding a lug to a recycled grid.

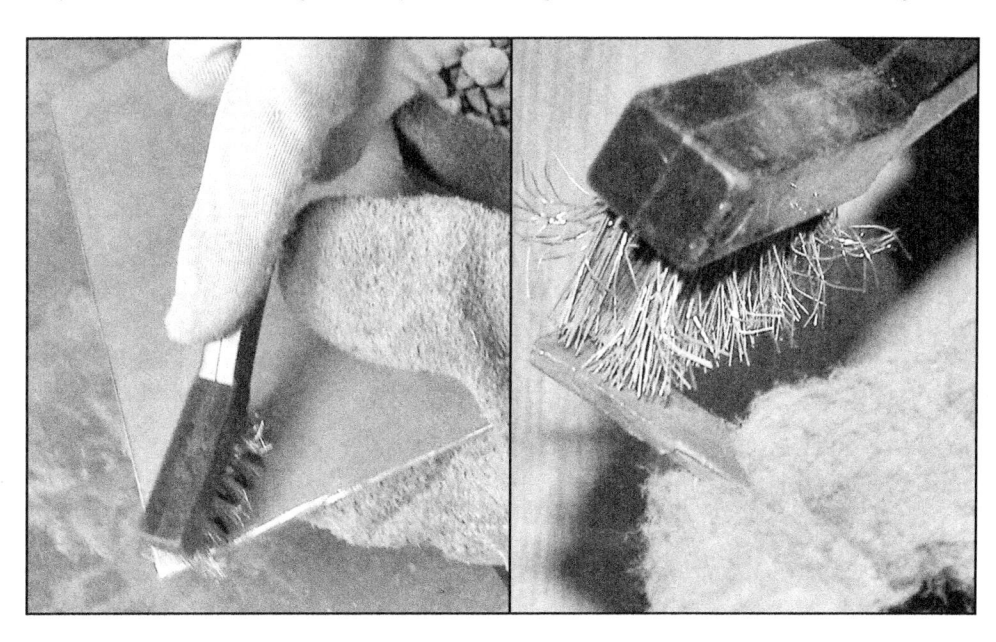

exact length you need. Make sure both pieces to be joined are cut straight and will fit against each other evenly. Scrape the plate at the spot that will be joined to the lug, scrape the lug and clean the welding wire. This will remove oxides that can interfere with an easy joining. Make sure the ends to be attached are straight and the parts abut each other evenly. You can use casting putty to hold the plate and lug so they do not move while burning.

You can apply some ruby fluid to clean and keep excess oxides from forming. This is not necessary. A good scraping of the surfaces that will be joined should suffice.

You can hold parts in place with casting putty.

Applying ruby fluid

To begin the weld, melt a bead from the lead wire and let it drop onto the beginning of the weld. Then, with the torch, melt the space just at the edge of the drop and the two pieces to be joined. You will be melting both lug edges and the drop into each other. Melt another drop and let it fall halfway on the previous melt and half onto the surface of the edges to be joined. Apply the torch to the edge of the bead and two surfaces and melt them into each other. If done correctly the bead will flow right into the edges and everything will melt into one piece. Continue this process across the lug.

When welding, always hold the torch in one hand and have the welding wire in the other, ready to apply if you create a burn hole and you need to

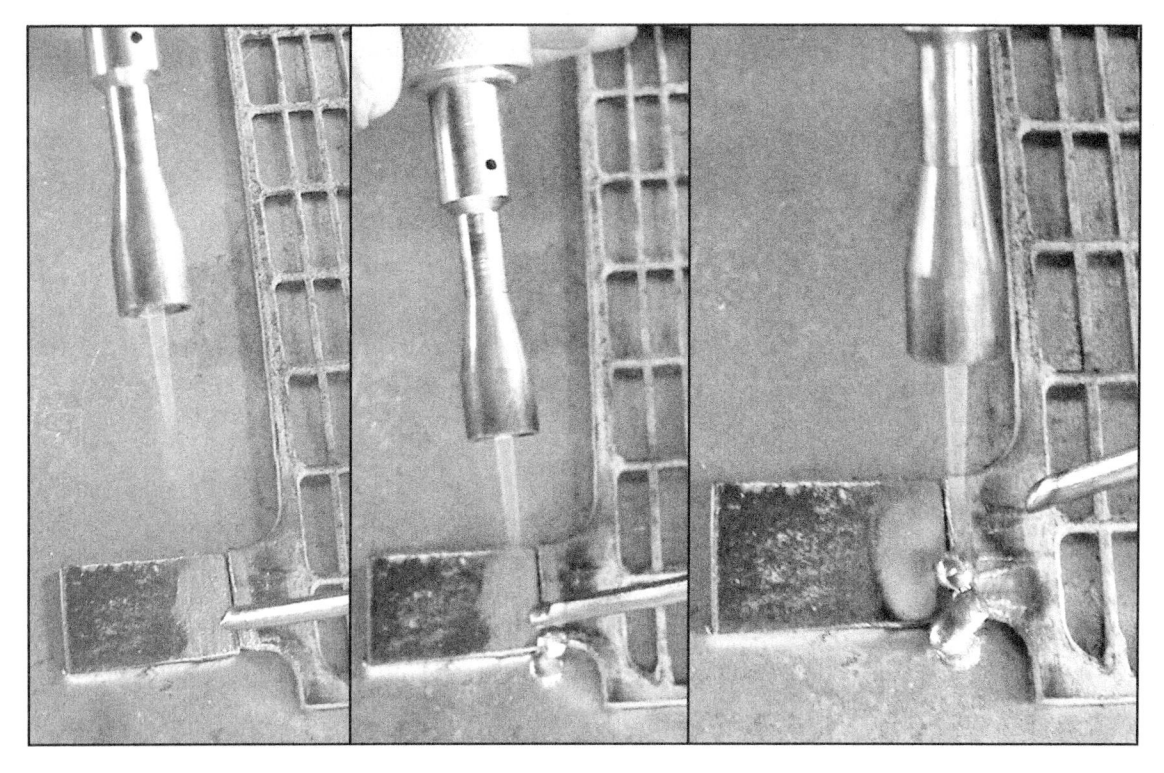

Burning Lugs onto Plates

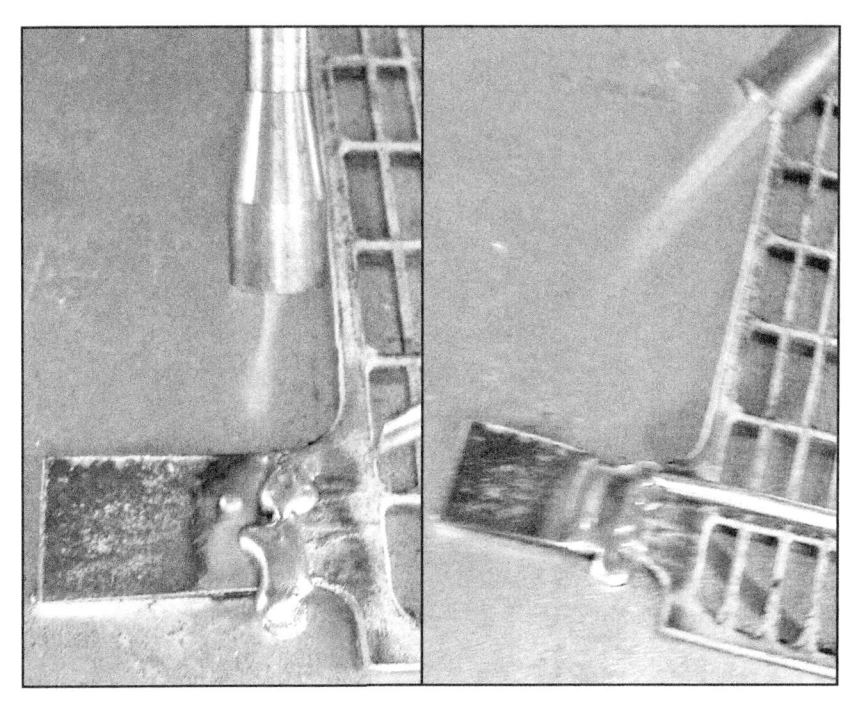

fill the void with lead. If this happens, melt some of the wire with the torch over the spot where the lead is separating to fill the space. This should not occur, but it will if you hold the torch on one spot too long. There is a little bit of timing to be learned in the process to get a good weld but you will pick it up quickly. If you have excess drip over, do not worry about it as you can cut and trim it off after the weld. Your welds may turn out a little bulky and you may have to file down the bulk if it will interfere with placement in the strap slots. Bulk and drip can be avoided by welding with the proper size wire. For small seams like these, wire ⅛" in diameter or under will do a good job. If you practice welding some seams with different sized wire you will quickly find the right diameter for your project.

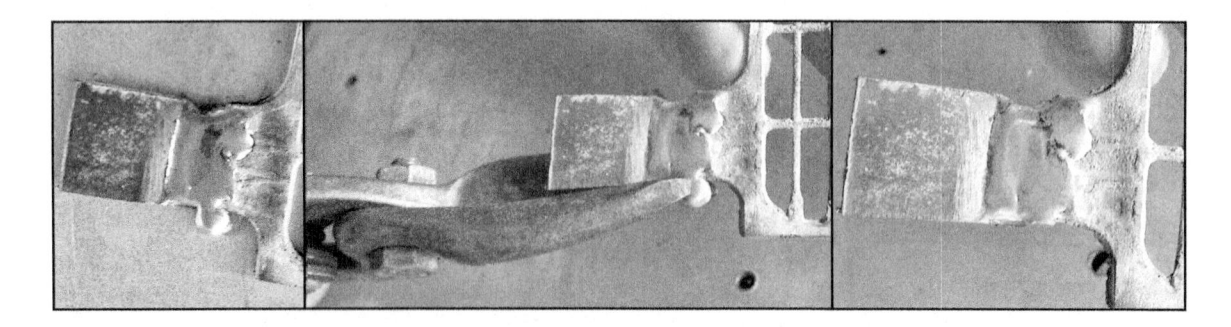

After welding, even the seam of the weld with metal shears and/or a file if necessary. All plates should be washed after working to remove contaminant particles.

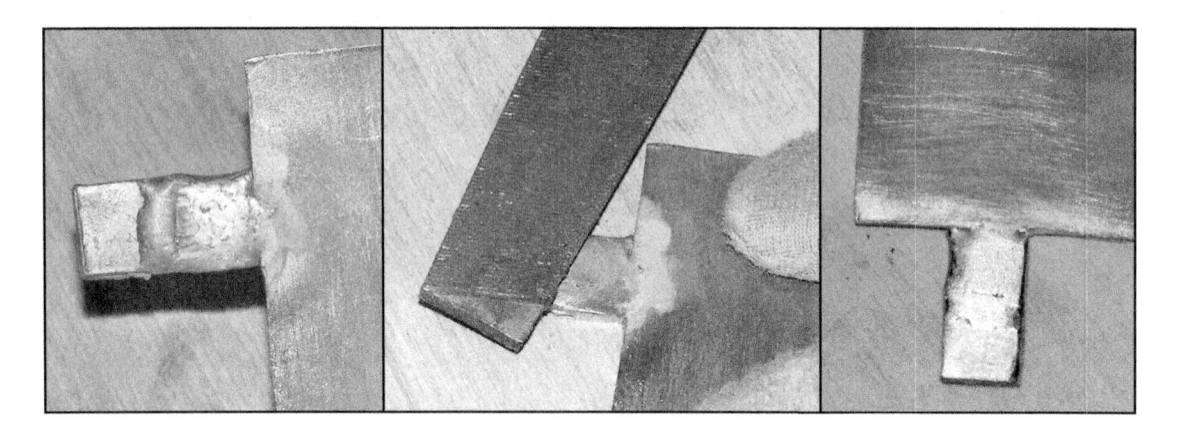

Laying in a melt

To lay in a melt to add a lug, or to repair a lug, lay the plate on the leveled burning rack, and build a dam to form the desired shape of the lug. Casting putty works well for this.

When the dam is set up, heat and melt the end of the existing lug remnant or the edge of the plate that will be joined to the lug. Then melt the lead wire into the form created by the dam. Like other lead burning

Burning Lugs onto Plates

operations some practice on scrap stock before working on actual plates is a good idea to get a feel for the nature of the process.

When laying in a melt be sure that the melted additional lead is not too thick. If it is, it must be filed down so that it will fit into the grooves of the strap for final assembly. Lugs do not have to the exact thickness of the grid lug remnant it is welded to, but the joint must be a good weld.

Lead and its alloys are very easy to work with and are very forgiving. It does not take much time to learn to do a good job.

Burning Plates into Groups

Set up components in the rack

To weld plates to form groups you need a plate burning rack. The base of the burning rack and the upper comb must be level so the melt created during the plate burning process remains even and does not shift to one side or another.

Before setting the plates in the rack for burning, check that the plates fit easily into both combs and the strap slots. If not, file as necessary until they fit.

Next, lightly file, scrape or steel brush the surfaces to be joined on the lugs, edges, corners and the slots in the straps to give a clean surface that will help in bonding the pieces together.

Burning Plates into Groups

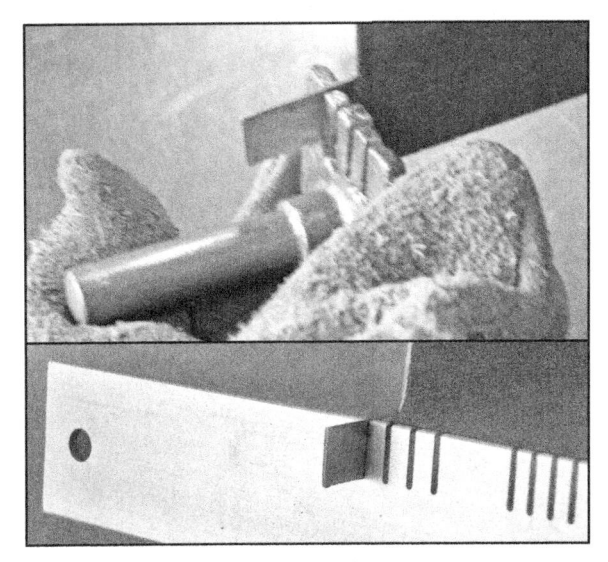

Place the plates in the slots in the upper comb. Slide the lower comb into position to hold the plates at the bottom.

Place the strap slots onto the protruding lugs on the upper comb and push the strap down onto the upper comb to fit as needed for burning.

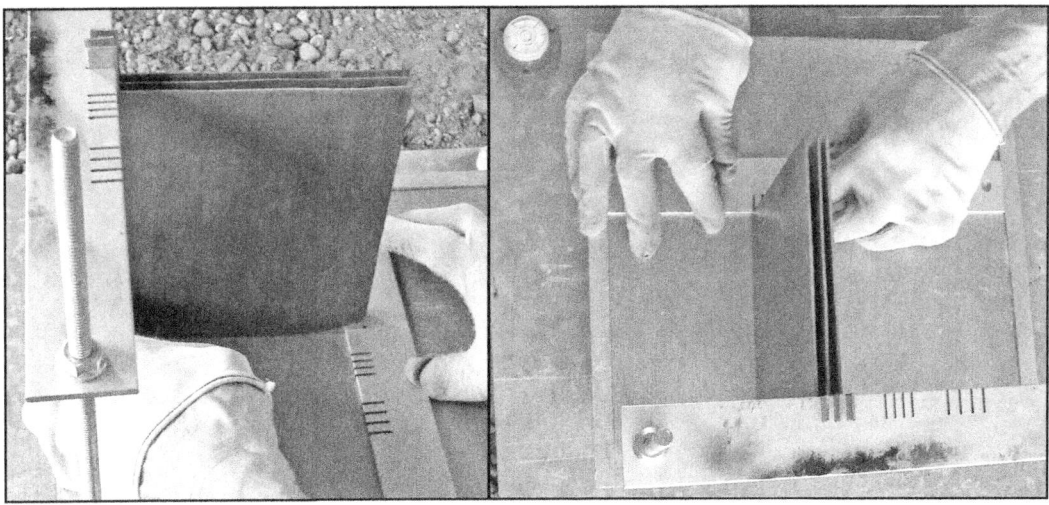

When the strap is slipped onto the lugs, squeeze the strap slightly onto the lugs for a tighter fit. This will also hold the strap slightly but firmly onto the lugs so the strap will be less likely to move while burning.

Prepare the components

Although not necessary, at this point you can brush a little ruby fluid flux onto the surfaces to be bonded. This may help to reduce oxide buildup and make welding faster. If there are pits that you will melt scraps into, you can also brush those with ruby fluid.

Burning Plates into Groups

Fill any voids or pits with scrap pieces. This is not necessary but can help to give more integrity to the structure.

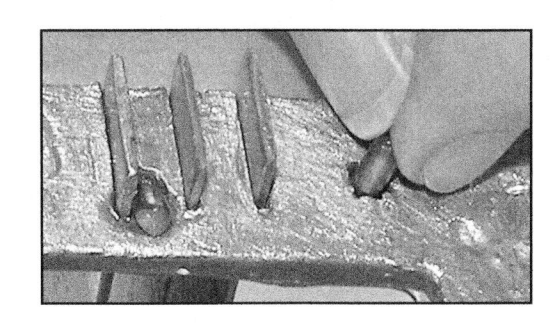

Set up heat shield cloth

For both post and internal intercell connectors, cut a piece of heat shielding fabric (fiberglass/vermiculite cloth or carbon cloth) to wrap around the post or intercell extension before you weld. This provides a barrier to the heat so that the posts and extensions do not melt during burning.

Cut a small piece of either material and wrap it around the extension or post so that it completely covers the area to be protected. You can fasten the cloth together with a paper clip. You can make a heat shield for each type of strap and then reuse it on each burn by simply slipping it on and off.

Set up the containment dam

A dam has to be set up around the perimeter and underneath the strap sp that no molten lead flows out during the burning process.

Dams can be constructed of metal or casting putty. I prefer casting putty. It can be used with uneven castings, can be applied to the underside of the comb, is easy to apply and sticks well to the surfaces. Although not necessary, I apply putty to the underside of the comb between and around the area of burning, where the lugs and the strap slots intersect.

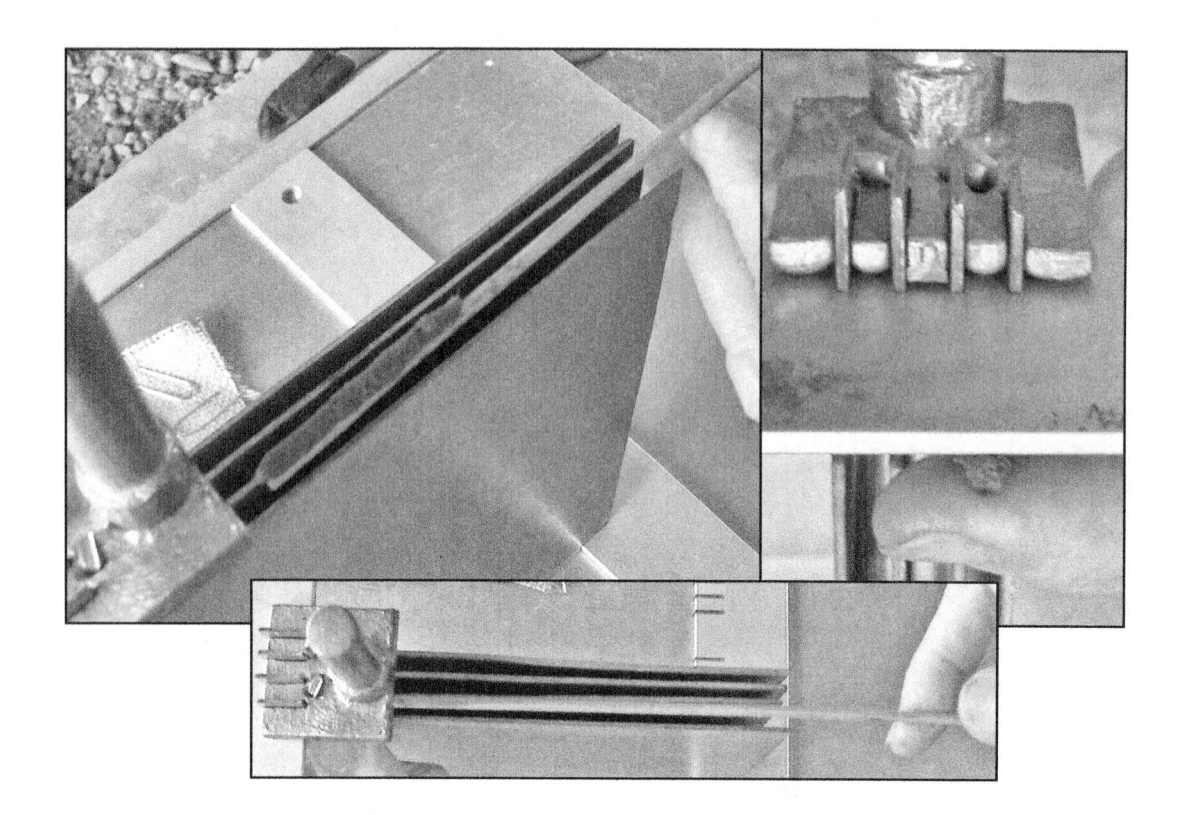

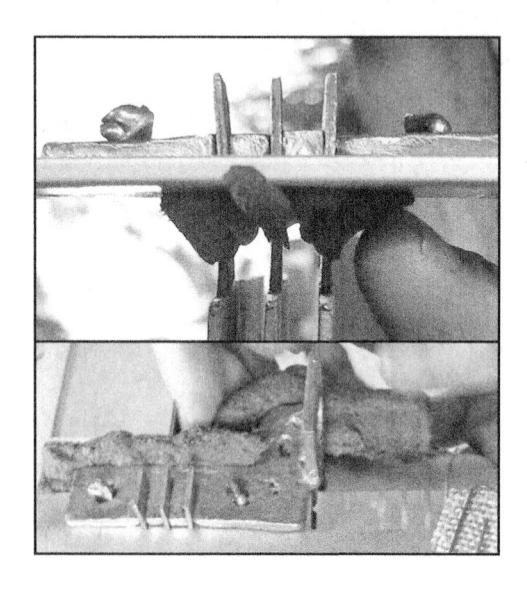

To apply the putty to the spaces between the plate lugs on the underside of the comb, I roll up a long thin piece of casting putty in my hand, balance it on the end of a BBQ stick, push it between the plate lugs and then push the stick upward to compress the putty against the plates and strap. This prevents the leaking of molten lead down onto the plates during the burn.

I also surround the outside lugs and slots underneath the strap with putty, completely surrounding the melt area.

Burning Plates into Groups

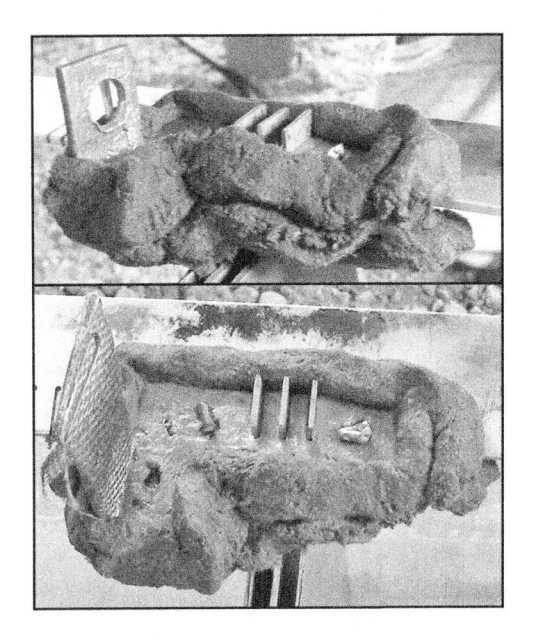

On the top, I build a dam to ensure containment of the melt as shown in the photos.

The Battery Builders Guide

Two groups ready to burn. Below, with a post strap; left, with an intercell connector strap.

Final check

Before burning, check the cloth heat shield to be sure there are no areas of the strap exposed that should be protected from the torch. Snug it back into place around the putty if necessary. Check everything for levelness, check the positions of the components and be sure that everything is firmly seated and ready for the burn. When you adjust or work with one part, often another will move out of line slightly, so it is always wise to check and recheck the setup before you light the torch.

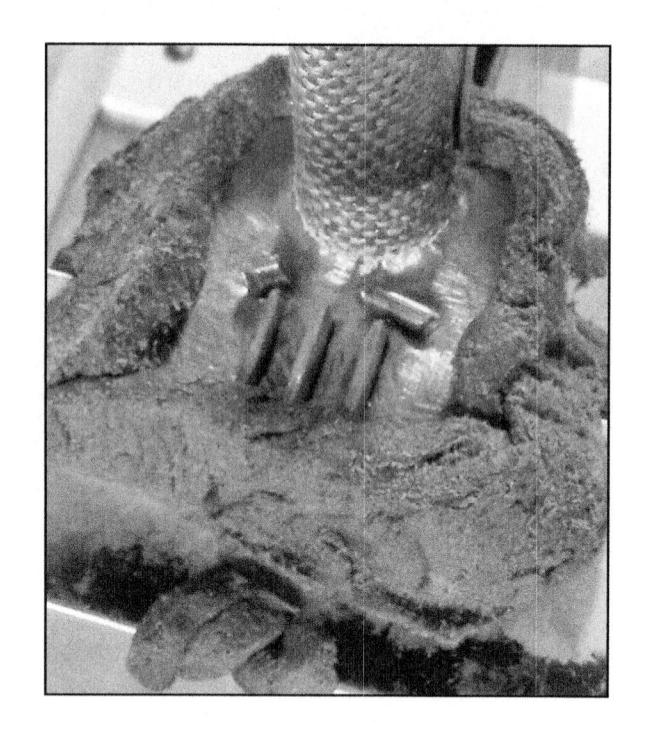

Burning Plates into Groups

The burn

Burning the lugs to the strap is pretty straightforward. Start by melting the protruding lug tops down to the surface of the strap and then melt the lugs and strap together in one continuous melt. If you need to add lead to voids or to even out the surface you can do so with the welding wire while burning.

The Battery Builders Guide

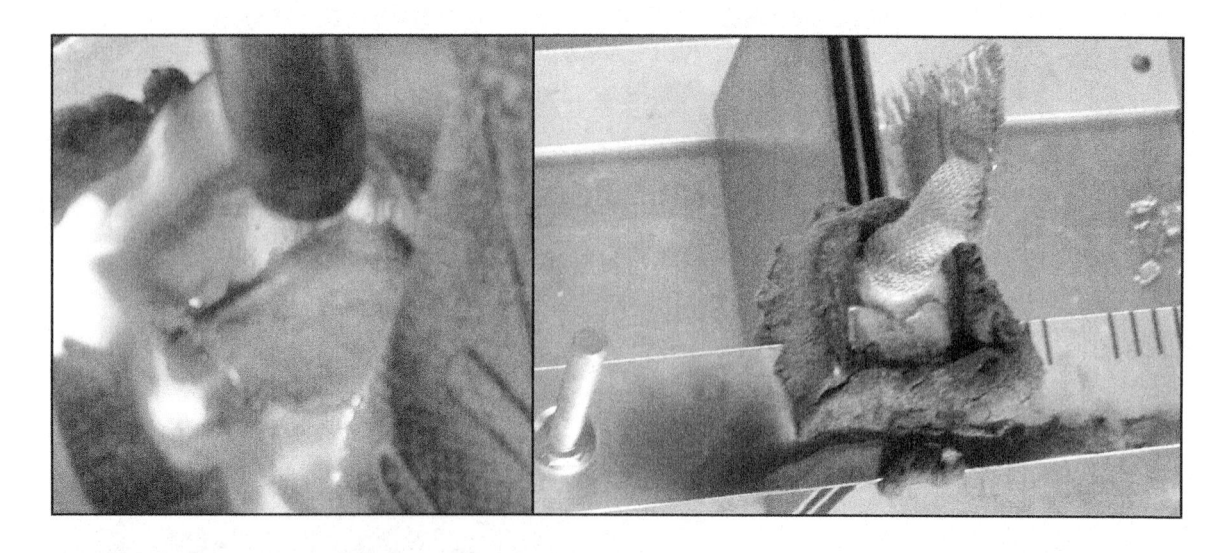

After the burn, let the strap cool down. Then, remove the casting putty, wipe off dross with a rag and steel brush the surfaces slightly to further clean. Follow this by another wipe with a rag.

Burning Plates into Groups

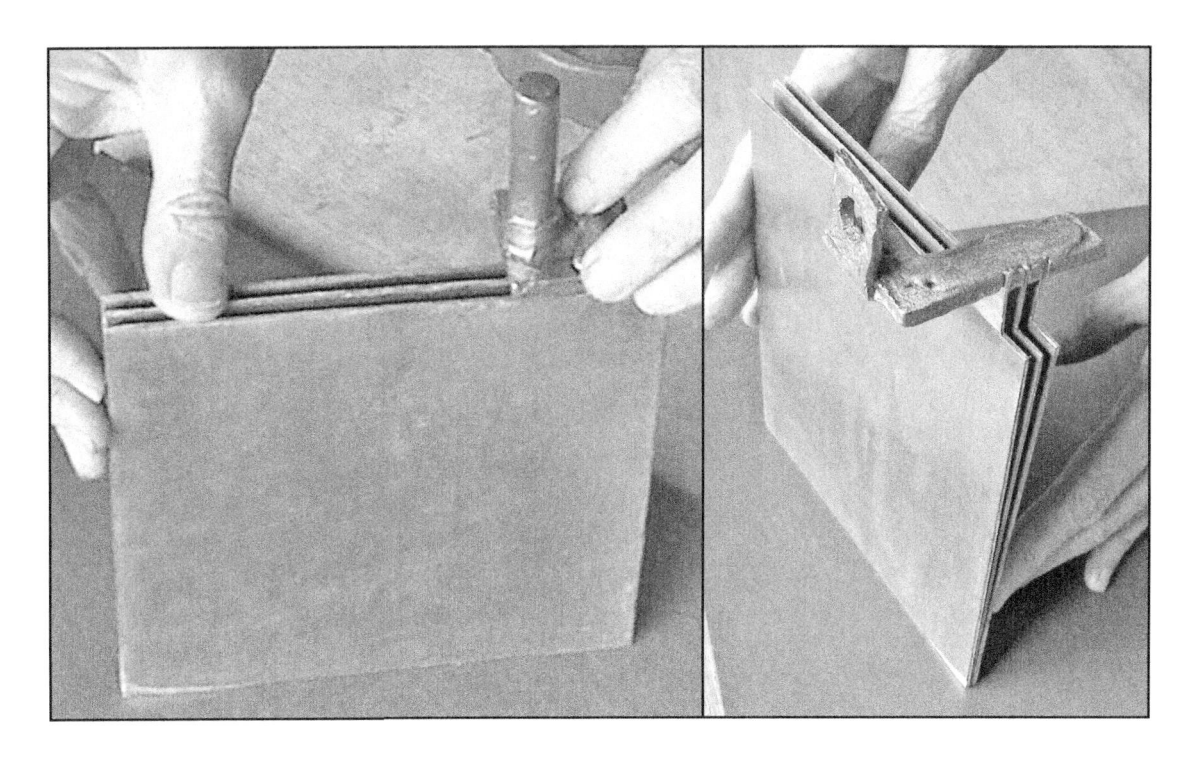

Remove the plates from the combs and inspect them thoroughly for pieces of putty that may be attached and for any melted lead runners or blobs that need to be scraped off. Make sure that there is no debris left on or under the strap, or in between the plates. You can use a cotton swab to remove any putty particles left between plates.

Carefully move and store the plates with support so that they can not fall over. I usually use a sturdy cardboard box.

The Battery Builders Guide

Battery Assembly

Plate Separators

Plate separators provide electrical insulation between the positive and negative plates in an element while at the same time allowing ion conduction from one plate to another.

Materials used for separators

Plate separators have to be porous, non-conductive, and acid and oxidation resistant. Antique batteries used perforated wood such as cedar, basswood, cypress, redwood and cherry that was boiled in an alkaline solution for about 24 hours to neutralize any organic acid which the wood contained. Other materials that have been used include cellulosic fiber mats, glass fiber mat, slotted or porous rubber, cloth, fiber mats, PVC, polypropylene, polyethylene.

Separator shape

Separators can be divided into three main types — leaf, pocket and sleeve — according to their shape and how they fit into the element. Leaf separators are simply flat sheets that are inserted between the plates. Pocket separators are double layered with the layers attached to each other along three sides, making a pocket that the plate slips into. Sleeve separators are double layered with the layers attached to each other along two opposite sides and are also slipped over the plate.

Ribbing

Separators do not have to be ribbed, but may be ribbed on both sides, or, as with most commercial batteries, just on the side that faces the positive plate. Separators in commercial automotive batteries are frequently webbed or ribbed to help somewhat to mechanically support the active mass on the plates while also keeping electrolyte space open between the plates.

Plate Separators

The value of ribbing is at best marginal in terms of battery performance. The webbing adds to electrical resistance and thus lowers performance. However, ribbing does have value in commercially made batteries that need to have plates packed tightly together to minimize space.

For the most part, a plain separator surface is fine. The plate lugs have enough flexibility for expansion of the space between the plates, which in turn allows gas to escape and electrolyte to constantly flow on the plate surfaces. This does require designing the elements to have a little space for this separation to occur. It will not work if your groups are crammed into the cells as tightly as commercial manufacturers do.

Mesh

Separators can also be made of polyethylene or polypropylene mesh with different opening sizes. Mesh is used when plates do not touch the separator. The mesh is a safeguard as electrical insulation.

Rigid mesh is inserted in rails between the plates. If one of the plates begins to buckle due to over discharge, the mesh will prevent it from touching other plates. Mesh by itself should not be used when plates are closely stacked as the mesh openings can be a bridge for conductivity when oxides fall from the plates during normal usage. This would, of course, short out the cell.

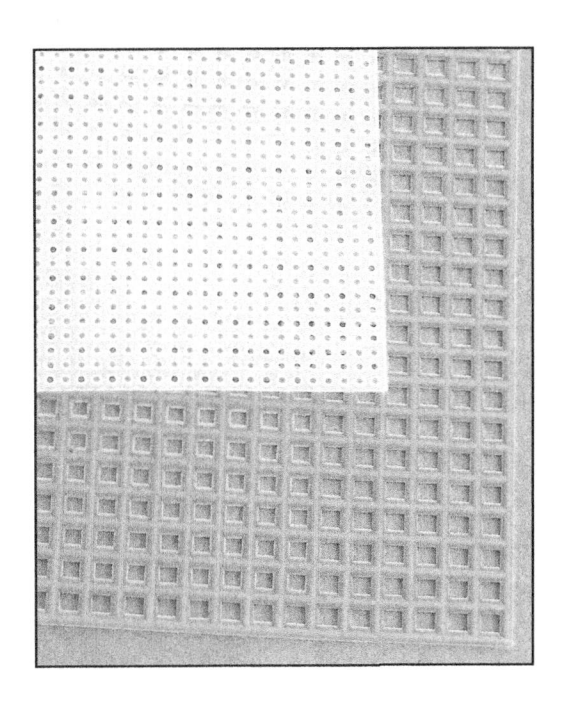

The solid structure of the mesh also does not allow gas to be released as readily as porous felt does. The mesh has to have a little distance from it to each plate in order to allow acid to circulate and debris to fall from the plate into the cell well.

Two types of mesh for separators.

Making sleeve separators

Separators are very easy to make. All you need is porous material such as polypropylene felt, some rulers, a rotary cutter or scissors, and a soldering iron or wood burner with a new clean tip.

Measure the plates. Each sleeve will be folded in half to fit over the plate, so for a 6"x 6" plate the felt would be at least 12"x 6" without the seam allowance. Add ¼" to the length and width to make a ⅛" seam allowance on the edges, and include extra length to cover the top of the plates to be sure they are insulated from one other. So, for a 6"x 6" plate, cut a piece of felt that is 6¼"x 12¼". If there are tabs on the bottoms of the plates, measure them and add to the length.

Once you have the measurements, a template can be made from thin plastic sign material that can be purchased from any local hardware store.

Cut the sign to the size needed, place it on the fabric and cut one piece. Before cutting the rest of the separators you need, fold and bond the first one and slip in a plate to test for fit to be sure the size is correct.

Cut out one separator for each positive plate. For two 6 volt batteries with three cells each and seven plates per cell (three positive, four negative), you will need eighteen pieces of felt to cover each of the positive plates.

Plate Separators

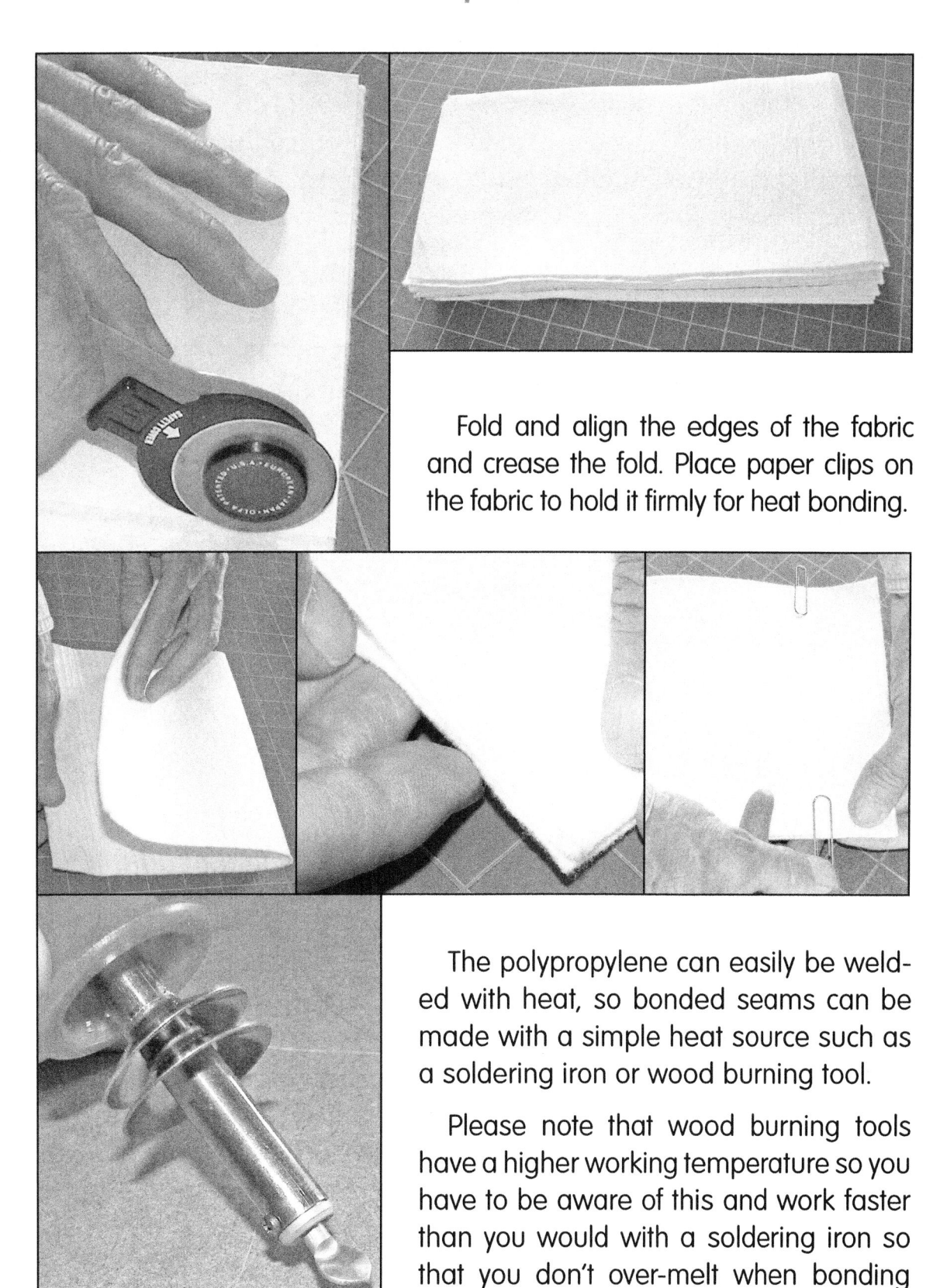

Fold and align the edges of the fabric and crease the fold. Place paper clips on the fabric to hold it firmly for heat bonding.

The polypropylene can easily be welded with heat, so bonded seams can be made with a simple heat source such as a soldering iron or wood burning tool.

Please note that wood burning tools have a higher working temperature so you have to be aware of this and work faster than you would with a soldering iron so that you don't over-melt when bonding the polypropylene. There are a variety of tips you can use with either a wood burning iron or soldering iron. Basically any type of tip will do but you may find it easier to work with one particular shape than another.

Set up your iron on a smooth work surface and position your felt so that you can easily move down the edges of the folded felt to bond the seams.

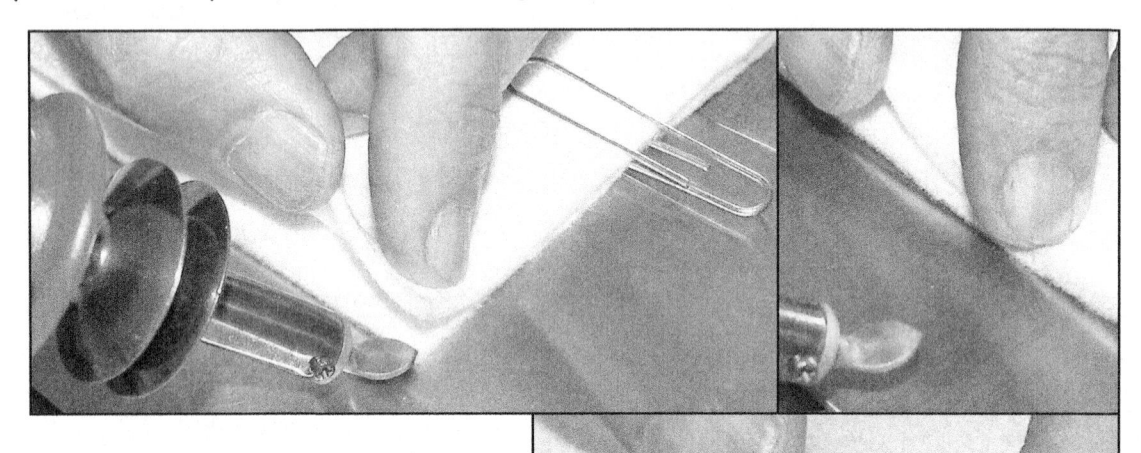

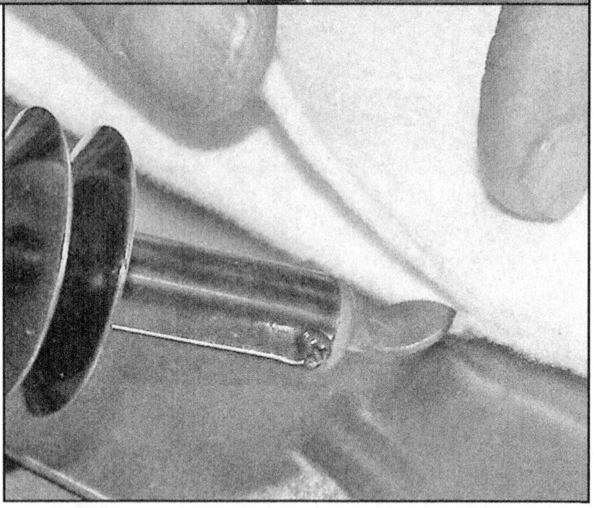

Position the hot tip of the iron between the two pieces of felt to be bonded as close to the edge as possible.

Press down on the top piece of felt so that it touches the top of the iron tip and pushes it down onto the bottom piece of felt. The iron tip will be sandwiched very briefly between the two pieces of felt and the inside edge will melt. As soon as the tip touches both edges of the felt it should be removed while at the same time you press the edges firmly against each other with your finger. Continue this process for the length of the sleeve.

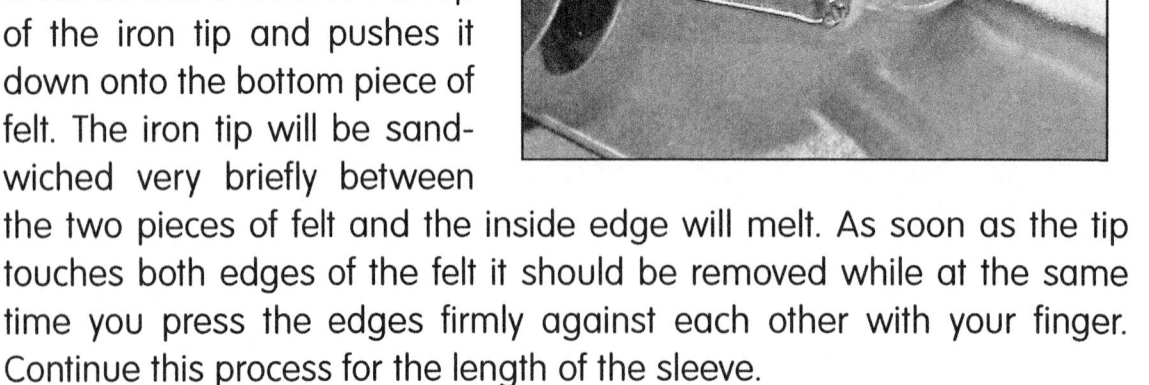

Plate Separators

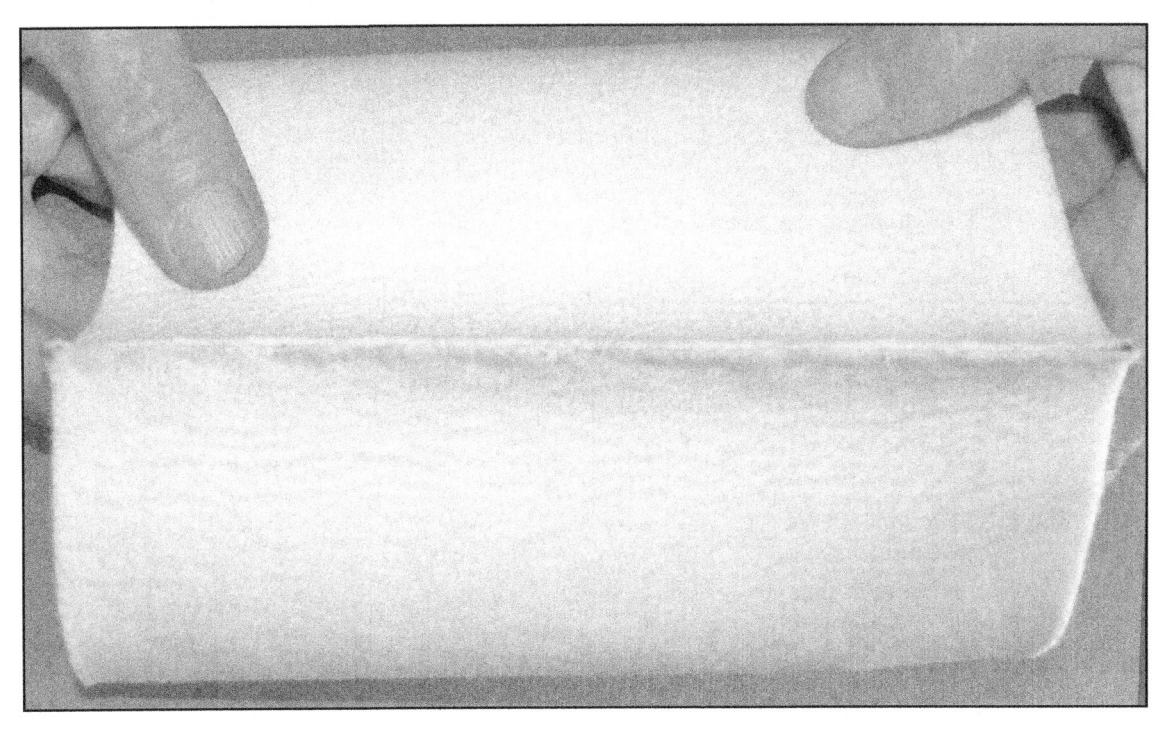

This technique should give you a perfectly bonded sleeve.

Slip in a plate to test the fit of each sleeve. The sleeve should fit on easily and should be a little loose rather than too tight.

The Battery Builders Guide

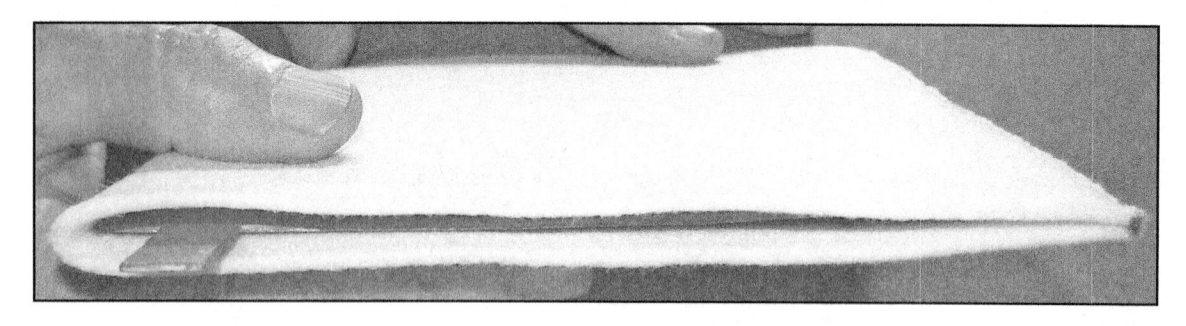

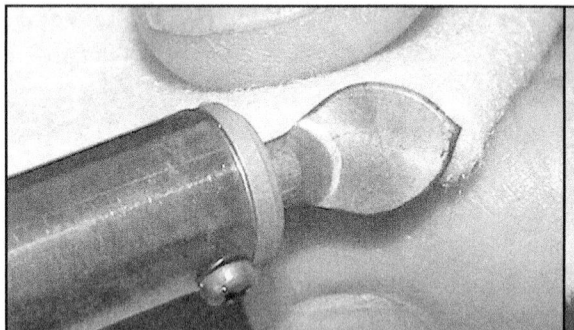

The next step is to bond the bottom corners. This will keep the felt from floating upward along the plates when they are immersed in electrolyte.

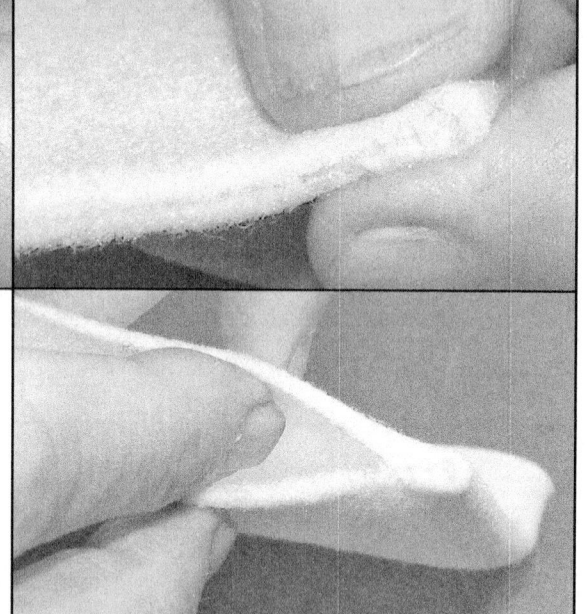

Battery Assembly

Final battery assembly requires envelopes, sleeve separators(**?), case and plate groups and shims if shims are used.

Inspect the case, each group, and the plate separators and shims to make sure they are clean and free of dirt and debris.

The Battery Builders Guide

Assemble the elements

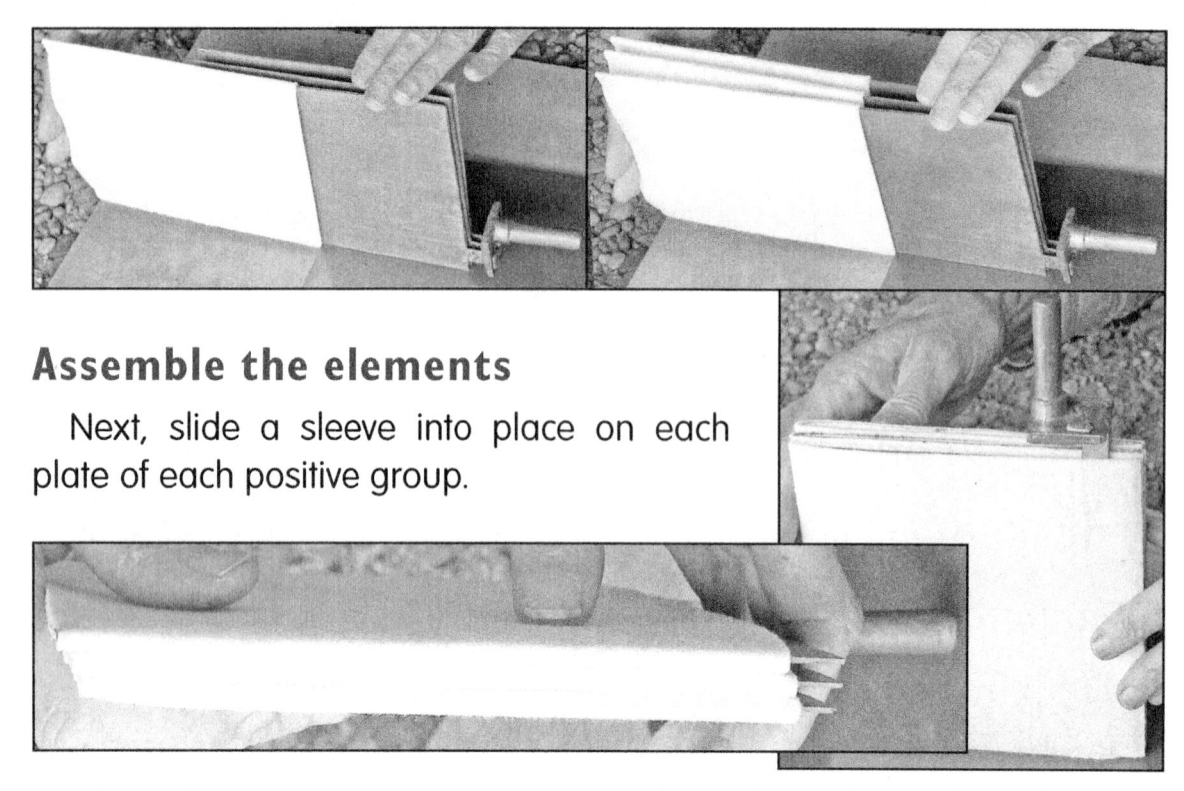

Next, slide a sleeve into place on each plate of each positive group.

After all the positive plates are covered, insert each positive group into a negative group to form an element.

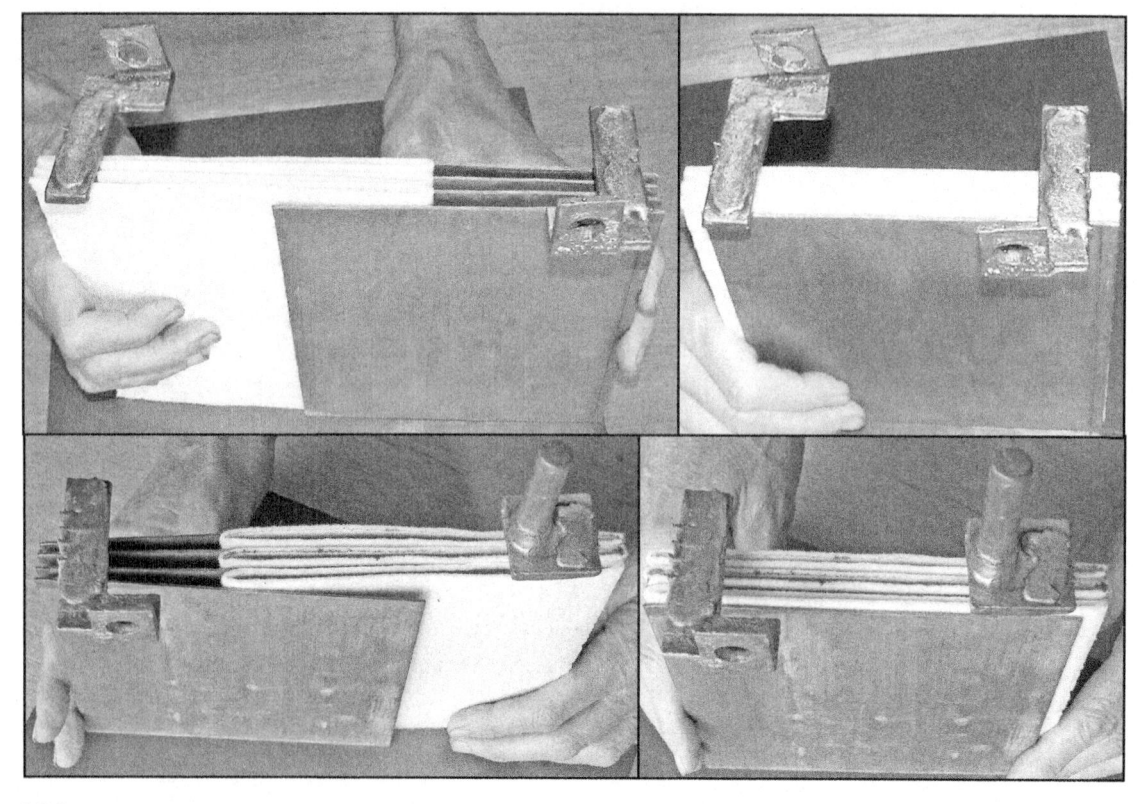

Battery Assembly

Fit and install the elements

Never lift groups or elements by the straps. Cut two pieces of string for each element and use the string to lower and raise elements into the cell wells for fitting and final placement. Lower the elements into the cell wells carefully.

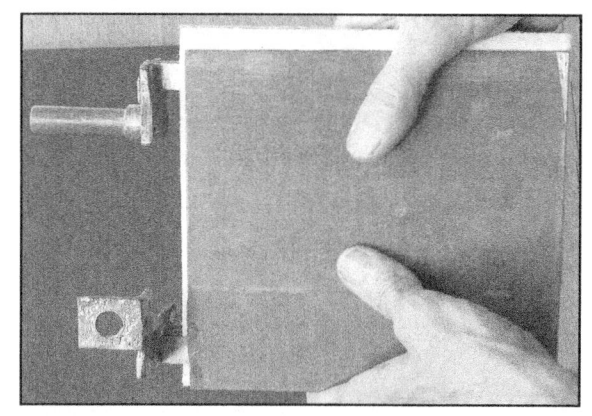

Keep the element balanced well with the string so that it does not fall and get ruined, and do not let the ends of the strings fall into the well.

Lay the strings to the side of the well. Check the alignment of the intercell holes in the case with the intercell strap connectors. If it is

199

off a bit you can bend and twist the element a bit. The plate lugs are quite flexible and this flexibility can make up for some minor inaccuracies.

If you are using shims, the alignment can be checked either before or after you place the shims in the cells.

When you have made the final placement of all the elements, check to be sure that each element is in the correct well. Once you are sure that everything is correct and aligned, pull the strings from the wells.

Adjust and align the holes in the intercell connectors with the holes in the cell walls and slip the threaded lead rods through the connectors and cell wall.

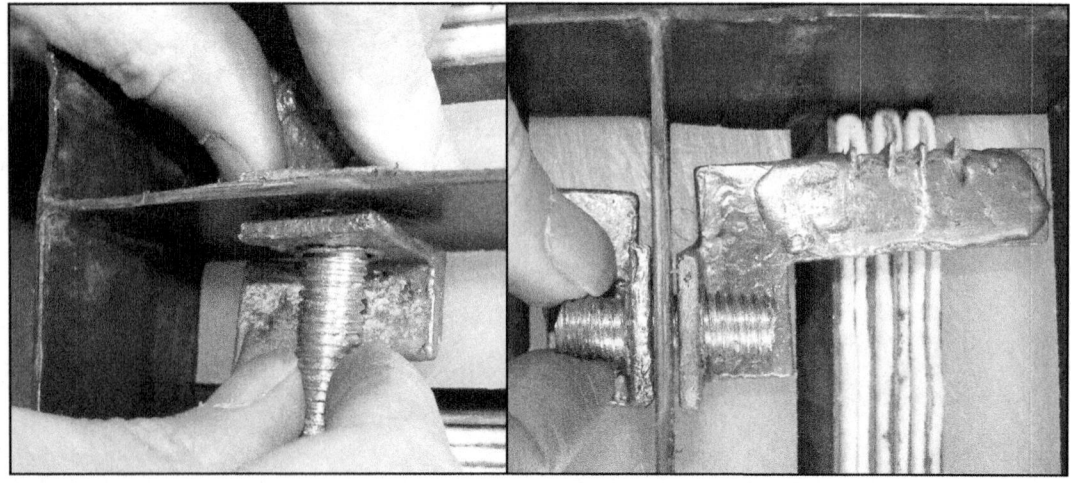

Battery Assembly

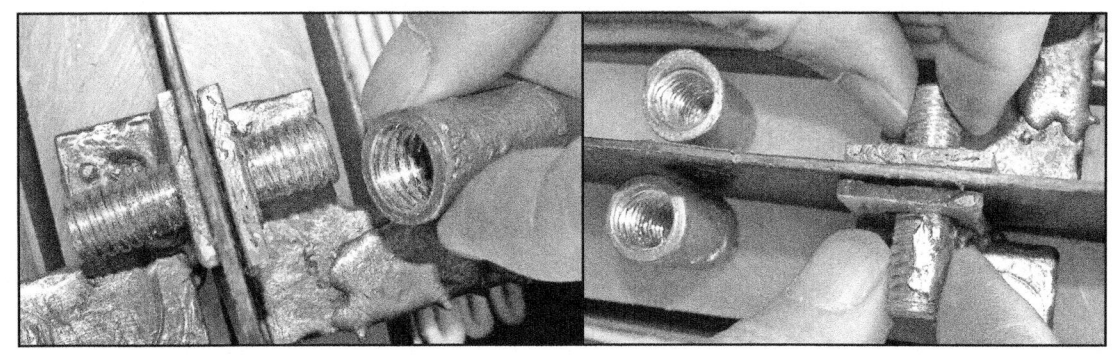

Test the pocket nuts on the threaded rods to see if the connection is tight. If they are a bit loose, add a thin lead connecting seal over the threads (see page 108). Press the seals onto the threads. Be careful not to drop the seals into the cell well, otherwise you will have to remove the element and retrieve the lost seal.

Apply Viton® caulk to the space between the case wall and the portion of the strap that sits against the well wall. Make sure you use plenty of caulk as it is important to make a good seal to keep one cell electrically isolated from each other.

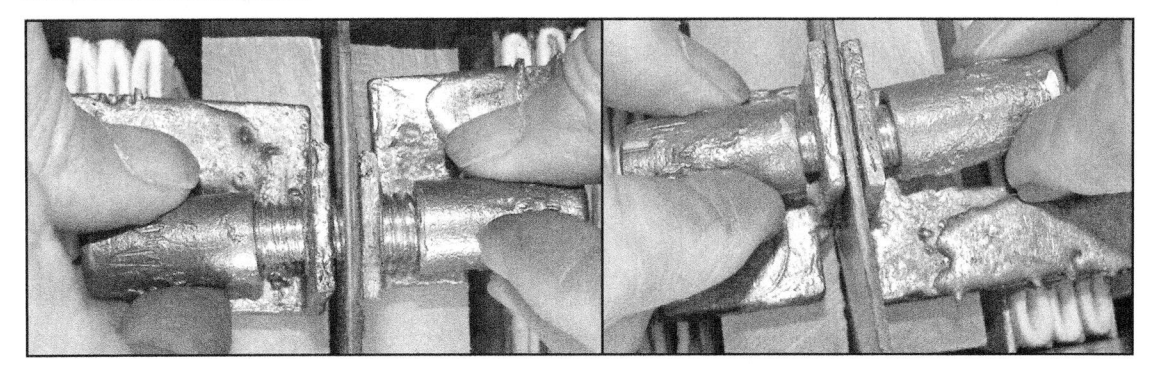

Hand tighten the pocket nuts on each intercell connector. Finish tightening by using pliers with moderate pressure to final fit. Do not overdo it with the pliers as the lead is soft. The idea is to firmly and tightly seal the nut against the strap connector without using excessive pressure. A loose connection can create resistance, which can create performance problems.

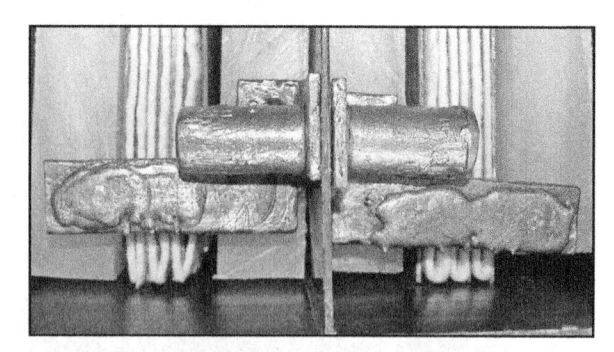

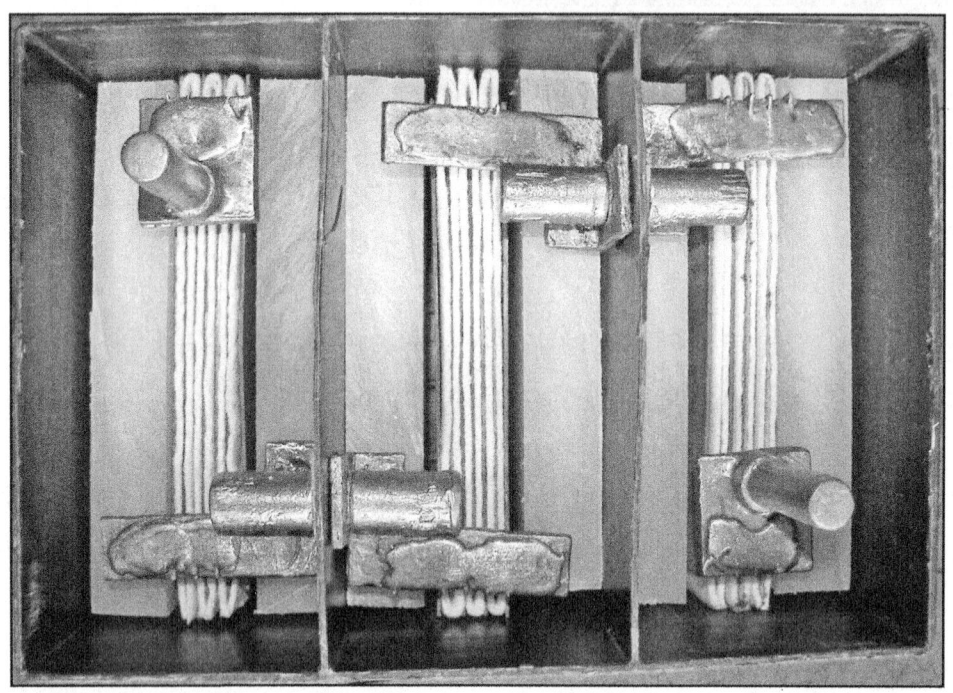

After tightening, apply a final coat of Viton® caulk around the edges of the intercell connectors to seal any gap between the cell walls and straps. Make sure to apply the Viton® all around the edges of the straps on the walls to ensure a very good seal.

Check for continuity between connected intercell connections, and between the positive and negative post. There should be continuity between intercell connectors that are connected to each other, and no continuity from positive to negative post. If there is continuity between the posts, there is a short and you need to find it. If there is no continuity between intercell connectors, you have a very loose connection that needs to be tightened with added connection seals or other means.

Battery Assembly

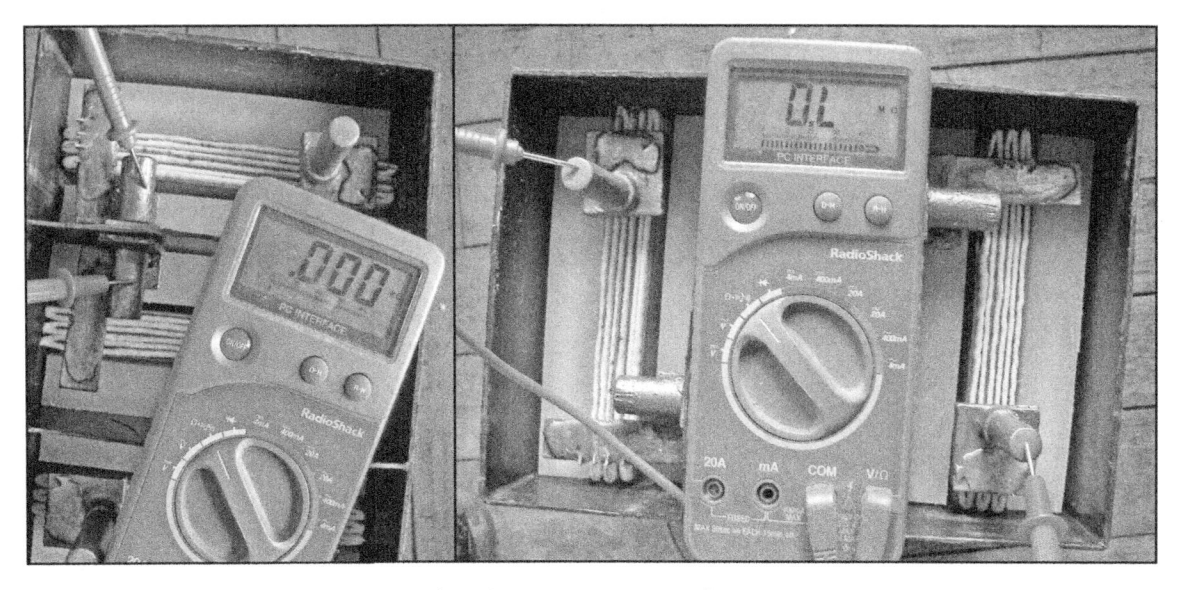

Above, check connections for continuity

Install the cover

Place the cover on the case to see how it fits over the terminals.

You may have to adjust and move the elements or terminals a bit to align.

Once the cover fits onto the terminals well, remove it and apply Viton® caulk to the top of the ridges inside the cover that will mate with the tops of the cell walls of the case. The Viton® will act as a seal. Apply epoxy to the inside side edges of the cover. This will firm the attachment to the case.

Place the cover onto the battery case and make sure it is fitting properly. The case top may have a little bit of spring in it. If so, put weights on top of the cover to hold it in place until the epoxy hardens. Let it harden for 24 hours.

Apply a final seam to the outside joint of the case and battery top with a final seal of polypropylene glue with a glue gun.

Finally, apply Viton® caulk around the terminal post bottoms where they meet the case for a good seal. Let the caulk dry for 24 hours.

Battery Assembly

Add the electrolyte

Set up acid dispensing system, remove caps from battery and fill the batteries to your designated fill level. Do not overfill. Fill each cell equally and cover the tops of the plates. When filling is complete, place and tighten the caps on the battery fill ports and get ready to charge the battery.

Charging and Forming

Plate Forming

Forming and charging basics

The ideal forming and initial charging voltage for both Plante and Fauré plates should be about 20% to 23% above the designated battery voltage, for instance a 12 volt battery should be charged at 14.4 volts to 14.76 volts, and a 6 volt battery should be charged at around 7.2 to 7.38 volts to avoid excessive gassing. The ideal charging current available should be about 10% to 13% of the rated amp-hour capacity of the battery. Thus, a 30 amp hour battery should be charged or formed at about 3 to 3.9 amps.

Designated vs. actual voltage

"Designated voltage" does not mean the actual voltage of the battery. Designated voltage is what a battery is named for its general output, ie, a 12 volt battery, or a 6 volt batter. Actual voltage is what a battery is outputting a given moment.

Forming for Plante plates

Forming Plante plates is a simple process. The cell, battery, or batteries are charged for a specific period of time, allowed to rest for several hours and then discharged for a specific period of time. This cycle is repeated 30 or more times, switching the polarity for each cycle. The purpose of this is not to charge the battery or cells per se, but to create a porous layer on the surface of the plates. The formation process electrochemically etches the plates and creates surfaces on them that will be porous enough to retain oxides on one plate and spongify the other plate. When the plates have developed a good surface, a final charge of about 72 hours completes the process of formation, and is the initial charge of the battery.

Forming cycles

The first cycle of forming is usually a charge for eight to twelve hours, followed by discharge for six to eight hours. After the first formation cycle, the polarity is reversed, and charged and discharged again for another eight to

Plate Forming

twelve hours each. Again the polarity is reversed, and the cycle of reversing polarity, charging and discharging is repeated over again about 30 times.

The battery will have designated positive and negative terminals. For the first formation cycle, connect the positive terminal of the battery charger to the positive terminal of the battery, and connect the negative terminal of the charger to the negative terminal of the battery. For the second cycle, connect the positive terminal of the charger to the negative terminal of the battery, and the negative terminal of the charger to the positive terminal of the battery. The polarity is alternated throughout the whole formation cycle in this way. If you do 30 cycles, for the last formation cycle, the negative terminal of the battery charger will be connected to the designated positive terminal of the battery; and the positive terminal of the charger will be connected to the designated negative terminal of the battery.

Then, to actually charge the cell or battery, connect the positive terminal of the charger with the designated positive of the battery and the negative terminal of the battery charger with the designated negative terminal of the battery for a final 72 hour charge. Following this, always charge the battery positive to positive, and negative to negative as you would for any normal charging operation.

Discharging

Discharging can be done with any load that does not require more than 10% of the battery or cell capacity. I often use 12 volt incandescent DC light bulbs. They give a convenient visual indicator of the discharge process. A screw-in base (bulb holder) can be used with a variety of bulbs with different watt ratings. Screw-in light fixtures can be found at any hardware store.

DC bulbs can be found at some hardware stores and most RV and marine suppliers. Common wattages available are 15 watt, 25 watt, and 50 watt. To get a general idea of the current draw of the bulb, use Ohm's law and divide the wattage by the voltage. Thus a 12 volt 15 watt bulb will use 1.25 amps, a 12 volt 25 watt bulb will use a little over 2 amps, and a 12 volt 50 watt bulb will use about 4.1 amps.

For example, for a 30 amp hour battery, one 15 watt and one 25 watt bulb connected in parallel will draw a little over 3 amps, which is near the appropriate rate of discharge for a 30 amp-hour battery. You can also discharge at lesser currents. I often do this for testing purposes and use a bulb rated at about half the amps rated for the battery.

There are automotive type bulbs in the 6 and 12 volt variety with a wide variety of amp ratings that could be used for discharging.

Batteries, although designed and rated for a certain amp-hour capacity, in fact usually deliver about 80% of their rated capacity. Also, batteries discharged at a higher rate (more amps) do not give as much current in total before they are discharged as does the same battery that is discharged at a slower rate.

A simple knife switch appropriately rated for the current needed can be set in the charge and discharge circuit as in the photo. The switch will disconnect the battery or cell from the charger and then connect it to the discharge load. This saves time spent connecting and disconnecting wires. The knife switch in the photo is rated at 25 amps, which is more than sufficient for the 6 amp charger and the 1.25 amp draw of the bulb used for discharge.

The initial formation and charge for Plante type batteries is only the beginning of the process. The performance of Plante type batteries improves as they are used. The current available increases with each cycle of charge and discharge. Throughout the entire service life of the Plante type battery the output characteristics improve, and in fact just before the plates are about to dissolve is when they perform the best.

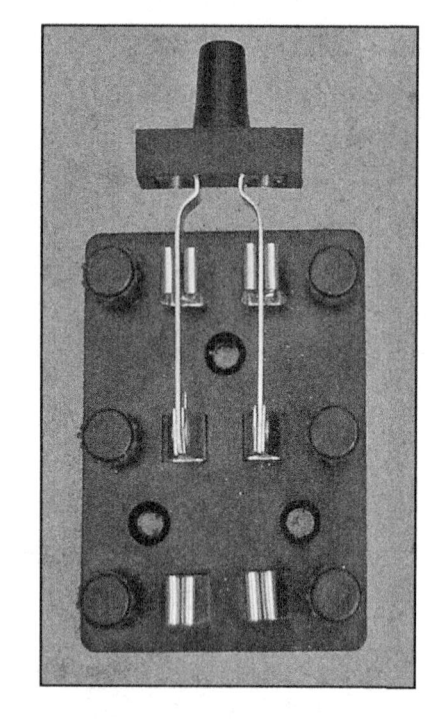

For Plante type plates another 30 to 40 regular charge discharge cycles

Plate Forming

are needed after the first 72 hour charge before the battery begins to stabilizes and begins to deliver its theoretical capacity. Although forming Plante plates is very slow, the result is pure lead plates that will last 10 to 20 years longer than pasted plates. Pasted plates performance deteriorates over time.

Initial charging for pasted plates

Pasted plates require from 48 to 72 hours of initial charging. They do not require polarity reversals. A pasted plate will stabilize and give its full capacity after about twenty cycles of charge and discharge. Batteries and cells made with pasted plates will have a service life of about 4 to 5 years. Since they do not require a lengthy formation period, they can be much more quickly manufactured than those with Plante plates.

Plante plates that have been formed. Negative, left and blue-gray; positive, right and reddish-brown. The color difference is a result of the formation process.

Equipment for Forming and Charging

Charging can either be accomplished using a grid powered charger or a renewable energy source such as photovoltaic panels or wind charger. Fast chargers should never be used for initial formation or charging.

Charging from the grid

The best charger to use is an industrial or plating rectifier with a blocking diode rated for the appropriate capacity, placed in series with the battery being charged. The blocking diode will prevent reverse current flow from the battery to the power supply, which will avoid catastrophic damage to the power supply. The most desirable plating rectifiers will have both variable current and variable voltage controls. A rectifier that provides up to 16 to 20 volts and up to 6 to 12 amps is a good choice, and can be used for a variety of battery sizes. They are more expensive but they allow precision control of the charging process. Used plating rectifiers can be found in the second hand market for a lot less than their original cost.

Manual chargers

The next best option and the most inexpensive is a basic manual charger that can be purchased at most any hardware store, automotive outlet, or large retail store. These chargers do not have pre-

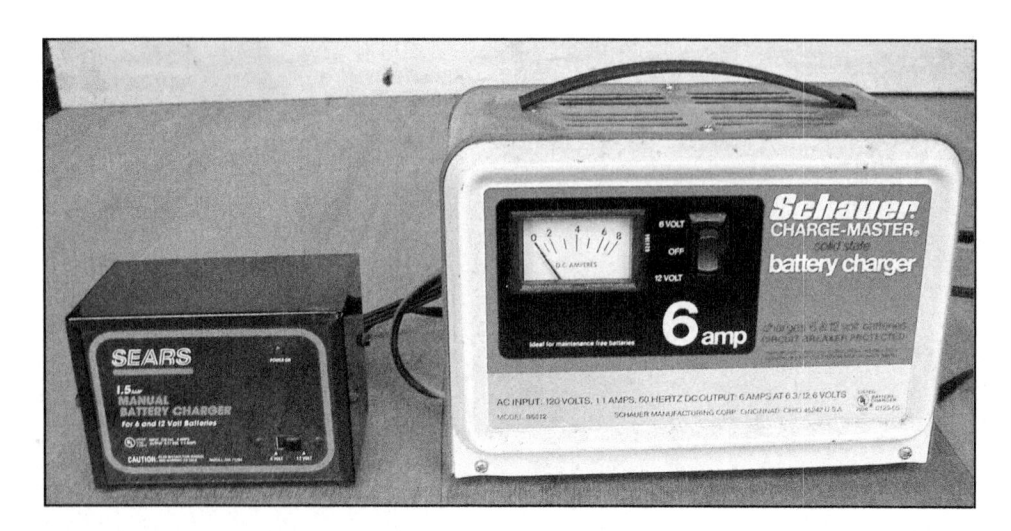

Equipment for Forming and Charging

cision control of voltage and current, but they work well for charging and forming, considering their very low cost.

Manual chargers come in a wide variety of current outputs. Motorcycle, lawn mower, motor vehicle chargers range from 1.5 amps through 12 amps. A common 12v/6v, 6 amp charger can be used to charge or form two 30 amp 6 volt batteries connected in series and charged as if they were a 12 volt battery. By charging the two batteries at once, you can use a charger that is rated for a higher current than you need.

Manual chargers do not adjust their input in response to the changing state of charge of the battery. Thus, they are called dumb chargers. Batteries being charged or formed with these chargers must be monitored for excessive gassing and overheating.

Smart chargers

Automatic chargers, also known as intelligent or smart chargers, cannot be used for initial charging and forming because most have a circuit that needs to sense at least 4 volts in the battery before it will turn on. Also, some have polarity protection, which makes it impossible to make the necessary polarity changes for forming.

Automatic chargers should be used for Plante type batteries after the initial forming and first 72 hour charge; and for pasted plate batteries after their initial charge. They will maintain the proper charging current and voltage according to the state of charge of the battery.

Charging multiple batteries

You can also purchase a gang charger which will charge a group of batteries at once, or purchase several inexpensive chargers to charge several batteries at once. If you are forming the plates in a Plante type battery, you can charge one battery, then rotate to the next battery while the other is discharging and so on.

The Battery Builders Guide

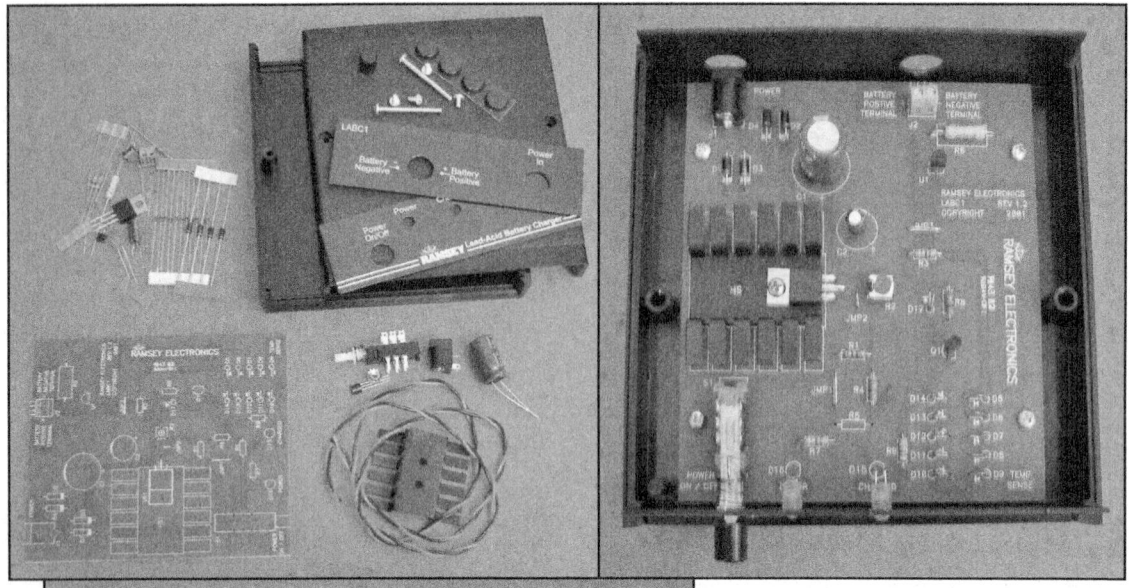

Ramsey charger for maintenance charging, but not for formation or initial charging.

Chargers for either grid or solar

Some chargers can be used both on grid power and input from solar, and some are more suited for smaller low amp hour batteries. These chargers come in kit form or can be purchased fully functional.

214

Equipment for Forming and Charging

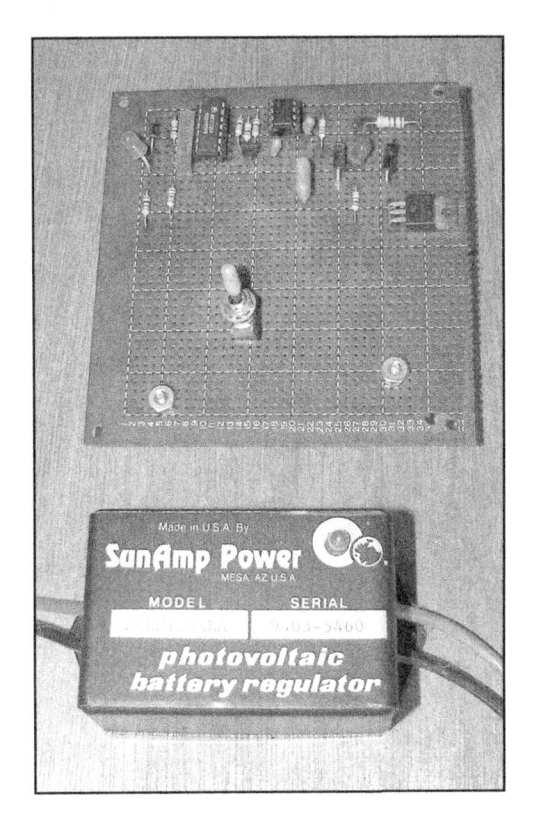

Solar chargers

If you have a photovoltaic system, you can use the diversion/dump load on the charging regulator to charge and form batteries after your PV system's existing batteries are fully charged. Depending on the amp hour capacity of the new batteries that you are forming or charging, you may need to insert a rheostat in the circuit to lower the current available to the batteries you are charging. In temperate climates this works best in the summertime, although there are some locations in these climes which get a lot of sun all through the year. In the summer, most battery banks are fully charged by late morning, leaving a lot of solar potential unused.

Homebuilt solar charge controller, top; commercial solar charge controller, bottom.

You can also set up one or more dedicated photovoltaic panels to serve as your forming and charging source with a simple charge controller. Photovoltaic panels come in a variety of current outputs.

Rheostat

If you set up a dedicated system be sure to use high current output panels. Two 12 volt 3 amp panels, connected in parallel will provide 12 volts and 6 amps for charging. As photovoltaic panels provide varying current output depending on the strength of the sun at any given moment, I would suggest at least a 6 amp and more preferably a 12 amp source.

The Battery Builders Guide

Completed solar charge control kit

You can make your own custom panels to output the exact current and voltage that you will need for your solar charger. Details on how to construct photovoltaic panels can be found in *Build Your Own Solar Panel*. You will need a charge controller with a dedicated system and you can find more information on these components in *Solar II*. Depending on the time of year I use either a grid powered charger or a solar charger, and sometimes both at the same time if the schedule is tight and I need to form and charge a lot of batteries.

The current from the charger can also be reduced by using DC light bulbs, fixed power resistors or a variable resistor such as a rheostat to limit the current to the batteries. Several rheostats connected in a certain fashion as in the unit in the photos (next page) will give control of both current and voltage output from either a solar source or commercial charger. The particular unit in the photo can handle up 80 amps and can be used to charge one battery or a gang of batteries. The circuit is basically a simple variable voltage divider and current limiter. If you construct such a unit you will need to add a blocking diode rated for the charging current and voltage, ie, if

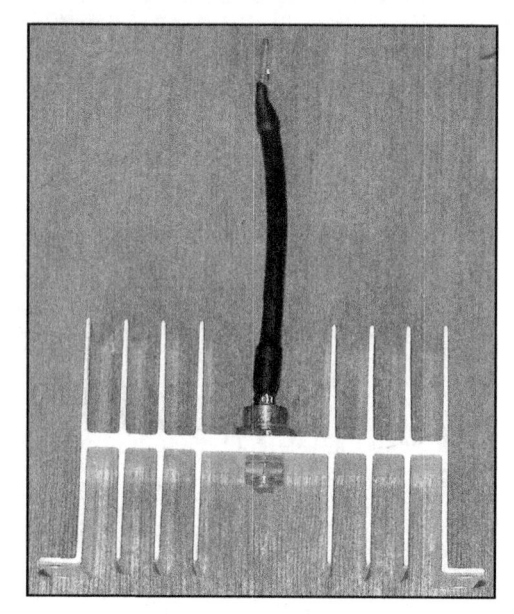

Blocking diode

Equipment for Forming and Charging

Home-built rheostat charge controller

your charging source delivers 6 amps at 16 volts, the blocking diode should be rated for more than the available current, and above the voltage from the power supply.

If you are interested in building a power supply rectifier, voltage divider, and current limiter there is a DC circuits tutorial listed on page 234 of the appendix.

Charging and Forming Procedure

Sequence for charging

1. **Disconnect.** Make sure the power source is disconnected, unplugged and switched off to the battery charger/rectifier circuit.

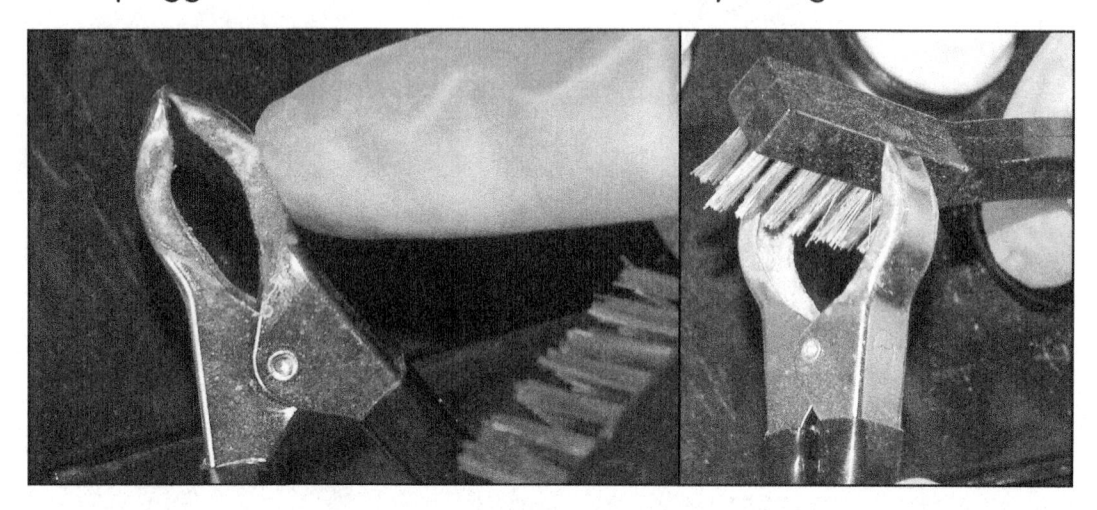

2. **Inspect and clean.** Make sure the battery terminals and connectors are clean. If not, brush them with a wet rag that has been dipped in a solution of baking soda and water. Then, brush the connectors and terminals with a steel brush. During charging and forming corrosion easily forms and this creates resistance, which means less current reaches the plates and heat is created as the energy is wasted. It is important to check the terminals and posts during the process and clean as needed, as it can make quite a difference

Equipment for Forming and Charging

in the efficiency of charging. Make it a point to clean all terminals and battery connectors after each cycle of charge and discharge. If you use a baking soda and water mix, do not let any of the solution get into the battery electrolyte — it will ruin it. I use a damp rag rather than brushing on the solution with a toothbrush. You can use a toothbrush after applying the damp rag to the terminals. A toothbrush helps to brush off the corrosion. Wipe all terminals and connectors with another rag wetted with distilled water only, and then dry with another rag after the corrosion is removed. You can then use a steel brush to scrape off any tough corrosion spots, and to clean surfaces for superior connections.

3. **Connect the leads.** Connect leads from charger or switch to battery terminals. Make sure leads are tight and in firm contact with the battery terminals. Be sure to check for correct polarity according to the operation you are performing. Please note that during the Plante plate forming process it is easy to forget what your last polarity position was. To avoid error, keep a written charging log of which lead is on what terminal.

4. **Check the charger settings.** Make sure the charger is correctly set to the proper voltage and current if applicable. If you have a DPDT switch in the charge/discharge circuit, flip the knife switch from the off position to the position that connects the charger with the battery.

5. **Turn on the charger.** Plug in charger, or switch on the power source and turn charger unit on if it has an on/off switch.

6. **Charge** for appropriate length of time.

7. **End the charging period.** Turn off power switch on charger and unplug charger. If you are using a DPDT switch in the charge/discharge circuit, throw it into the off position.

8. **Disconnect.** Remove clips from charger to battery from battery terminals. If you are initial charging pasted plate batteries, you are done.

9. **Let the battery rest.** If you are forming Plante type plates, let the battery rest for two hours and immediately proceed to the discharge sequence.

Discharging sequence for forming Plante type plates

1. **Turn off** all switches to the load.

2. **Connect** clip leads to battery terminals from load switch

3. **Throw load switch on** to commence discharge

4. **Turn switch off** to load when discharge period is completed,

5. **Disconnect** leads from battery terminals.

This completes one cycle of charge and discharge. Proceed to the next cycle as described on the following page.

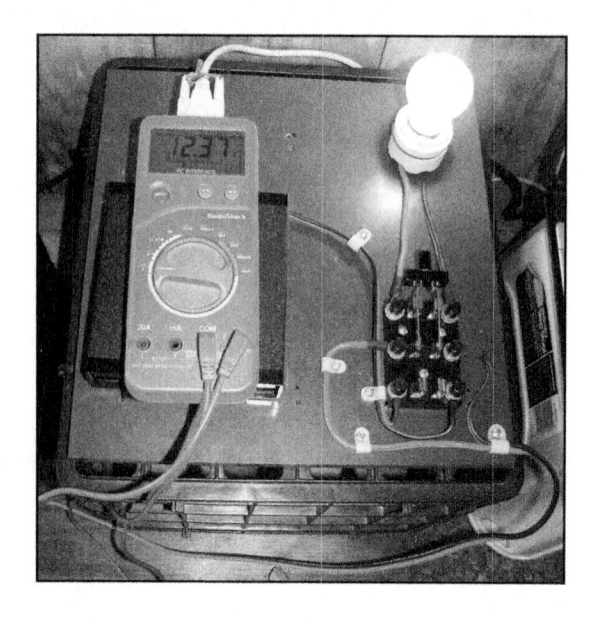

Equipment for Forming and Charging

Next cycle for forming

1. To begin the next cycle, **reconnect the leads to the load and perform a quick discharge** for about 1 minute or so. This will drain any residual bounceback voltage remaining after initial discharge.

2. **Disconnect the load leads and connect charger leads.** Connect the leads from the charger to the battery terminals in reverse of the polarity of the previous cycle of charge and discharge.

3. **Immediately proceed with another cycle of charge and discharge** (beginning page 220).

4. **Repeat the cycles of charge and discharge** a minimum of 30 times.

Final charge

When the 30 forming cycles have been completed, the battery is ready for a final charge of 72 hours.

Safety during forming and charging

The combination of gases generated during charging — oxygen and hydrogen — is highly flammable and will explode with the addition of an ignition source. You must make absolutely certain to keep all sources of ignition away from the batteries. Never smoke near charging batteries. There should be no open flames, or any equipment that can arc or spark anywhere near the batteries.

Keep the battery temperature below 110°F to avoid thermal runaway and to avoid excessive gassing, both of which could cause an explosion.

Charging or forming should only be done in a very well ventilated area. Outdoors is the best. The next best choice is to build a battery box with a small vent fan. The construction of such a box is discussed in my book *Solar II*.

Gassing

Light to moderate gassing is acceptable and desirable while charging or forming. The gassing circulates the electrolyte and breaks up stratification that occurs from the sulfuric acid being more dense than water.

Excessive heavy gassing, however, is not acceptable unless you are performing an equalization charge on the batteries. Equalization charges

Each cell in a battery should have the same output, but through continuing use, the cells' output may change at unequal rates. An equalization charge is performed on batteries at about 15.5 volts for several hours to maintain as near equal output from each cell as possible during general service use.

The maintenance schedule for equalization depends upon how much the batteries are used. If they are used heavily during general service, they should be equalized more frequently.

An equalization charge of about one hour can be performed on newly formed batteries, after batteries have had their initial charge. Batteries should only be equalized after they are fully charged in the normal fashion.

After equalization, always check electrolyte levels.

Electrolyte levels

Correct electrolyte levels need to be maintained during the initial charge or formation, and in subsequent general service use. Some distilled water is lost in the charging process as the water of the electrolyte solution is broken into its constituent gases, oxygen and hydrogen.

For Plante batteries, the electrolyte level should be checked after each cycle of charge and discharge in the formation process and at intervals during the last 72 hour charge. When charging pasted plates, the level should be checked at intervals during initial charging process.

The tops of the battery plates must be covered with electrolyte at all times. If the tops of the plates are exposed to air you can loose battery

Equipment for Forming and Charging

capacity for the rest of the battery life. If the electrolyte level is down below your preset fill level, add distilled water only. Do not add more electrolyte.

Only check the fill level after charging. During discharge, the sulfuric acid forms sulfates on the plates thus reducing the amount of sulfuric acid and thus the volume of the electrolyte. When the battery is recharged the sulfates are put back into solution and thus the volume of electrolyte is increased. If you check the electrolyte level after discharge and fill to level, you will find that when the battery is recharged, the electrolyte solution rises above the fill level and thus out through the top of the battery which is something you do not want. Do not use any type of water other than distilled. Do not overfill your cells. It is better to under-fill a little as long as your plates are covered. Maintain the same fill level for all cells.

Wiring and Connectors

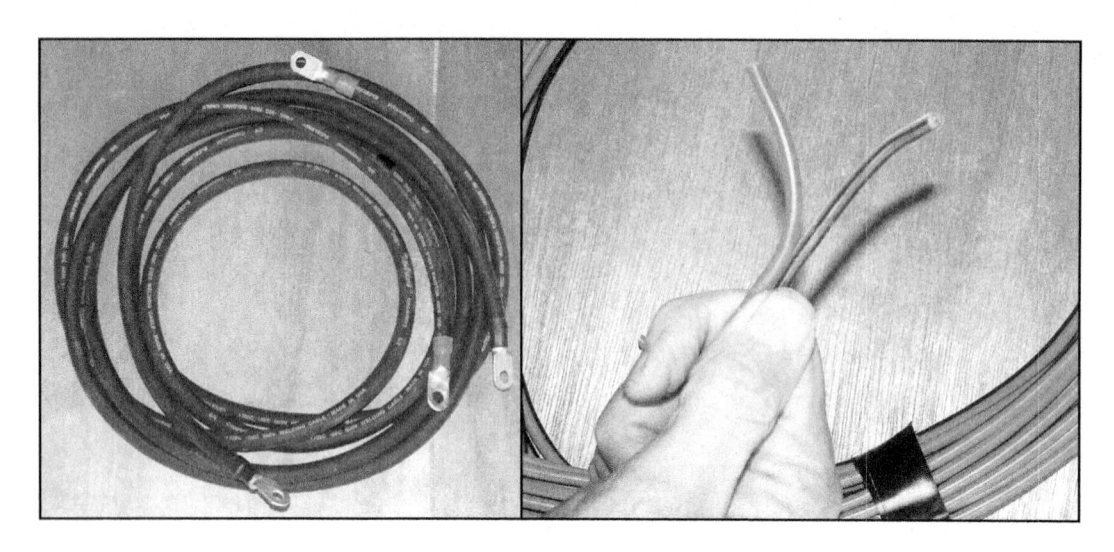

All wiring and connectors must be rated for the current supplied. Most grid chargers come with appropriately rated wire and clips. If you use any additional circuits, you must provide the correct wire size to carry the current being supplying. The same goes for connectors. For additional circuits, you will also need to insert a blocking diode and fuse of appropriate rating.

For inter-battery connects when charging in series, use appropriate battery cable or wire size that is rated well above the current input and the discharge output. All wiring and connectors should be rated at about twice any current input and output of the battery charging and discharging system.

The larger the diameter of the wire, the more current (amps) it can carry. It is important to size the wires you use throughout your system to match the needed current flow.

The amount of current that a conductor can safely carry is termed its ampacity. The ampacity rating of a wire is based on a number of factors such as the diameter of the wire, the type of insulation that covers the wire, the temperature of its surroundings (ambient temperature); and how close the wire is to other conductors that also

Wiring and Connectors

generate heat during conduction of current, and the nature of the raceway or run which the wire is enclosed in as this can contain and allow heat buildup.

Diameter of wire is an important factor in what is termed voltage drop. The longer the run of wire, the more resistance is encountered. This causes a loss in voltage. The resistance can be reduced by using larger diameter wire. Voltage drop is an important factor in any long wire run, but is really significant when charging or forming batteries from a photovoltaic panel with a long wire run.

AWG Ohms	Gauge per ft.	Ampacity
0000	0.0000490	312.0
00	0.0000779	220.0
2	0.0001563	131.0
6	0.0003951	65.2
8	0.0006282	46.1
10	0.0009989	32.5
12	0.0015880	23.0
14	0.0025250	16.2
18	0.0063850	8.1

Ampacity table

In general, using wire with a larger diameter than necessary is a good idea. At the very least, all wire and connectors in the system must be rated to adequately and safely carry the amperage needed. They should be also sized to minimize voltage drop as much as possible. The conductors also need to be able to conduct adequately and safely in the environment in which they are used.

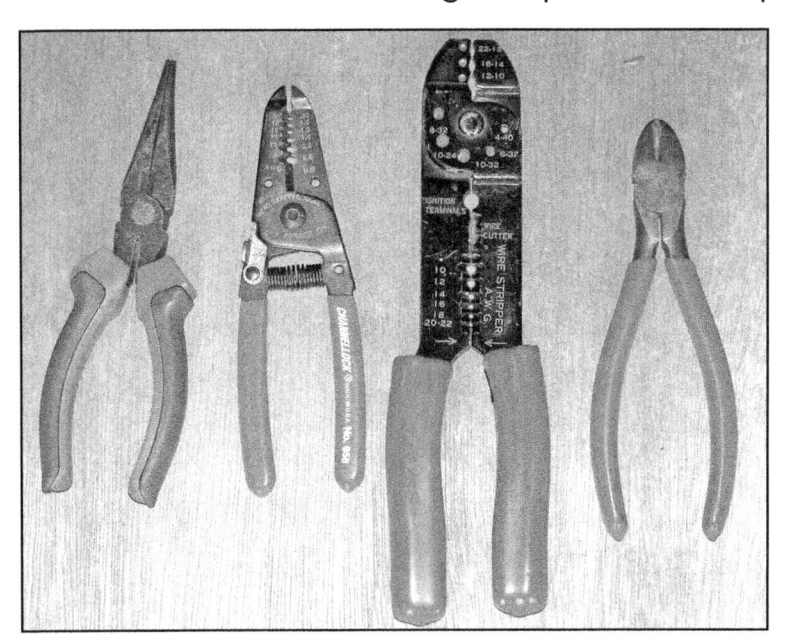

Make sure you have the appropriate tools to do good wiring, such as a wire stripper, cutter, wire terminal crimper and small pliers.

225

Always disconnect the charger from the power source before removing battery leads on batteries. This will avoid sparking and thus a possibility of explosion.

Formulas for working with batteries

Most electrical calculations that you will need to perform for batteries will be related to the relationship between volts, amperes and resistance. To perform these calculations, apply what is known as Ohm's law.

Formulas related to Ohm's law use:

	Symbol	Measured in
Current	**I**	amperes
Resistance	**R**	ohms
Voltage	**E**	volts
Power	**P**	watts

To find the resistance of a circuit or wire run when you know the voltage and current   **R = E/I**

To find the current in a circuit when you know the voltage and resistance   **I = E/R**

To find the voltage in a circuit when you know the current and resistance   **E = IR**

To find the power output of a circuit in watts   **P = EI**

Wiring and Connectors

Tests, performance characteristics and statistics

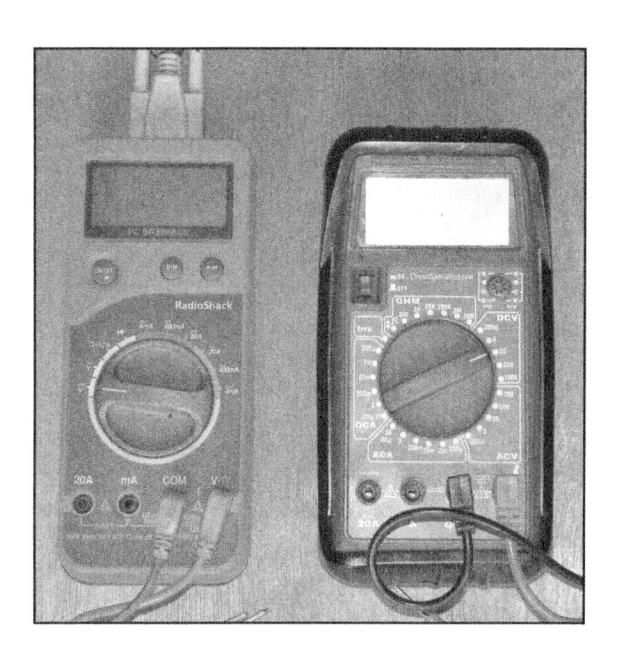

During the formation of Plante plates, it is very helpful to observe and collect information about the charge and discharge characteristics of the battery or cell. You can either keep hard logs — pen and paper notations of multimeter readings — or soft logs. To keep a soft log, you need a multimeter with analog to digital conversion that can also record and display on a computer screen; or a direct analog digital conversion unit with multimeter on PC that can display and record. The data logger can be either hard wired or a wireless system.

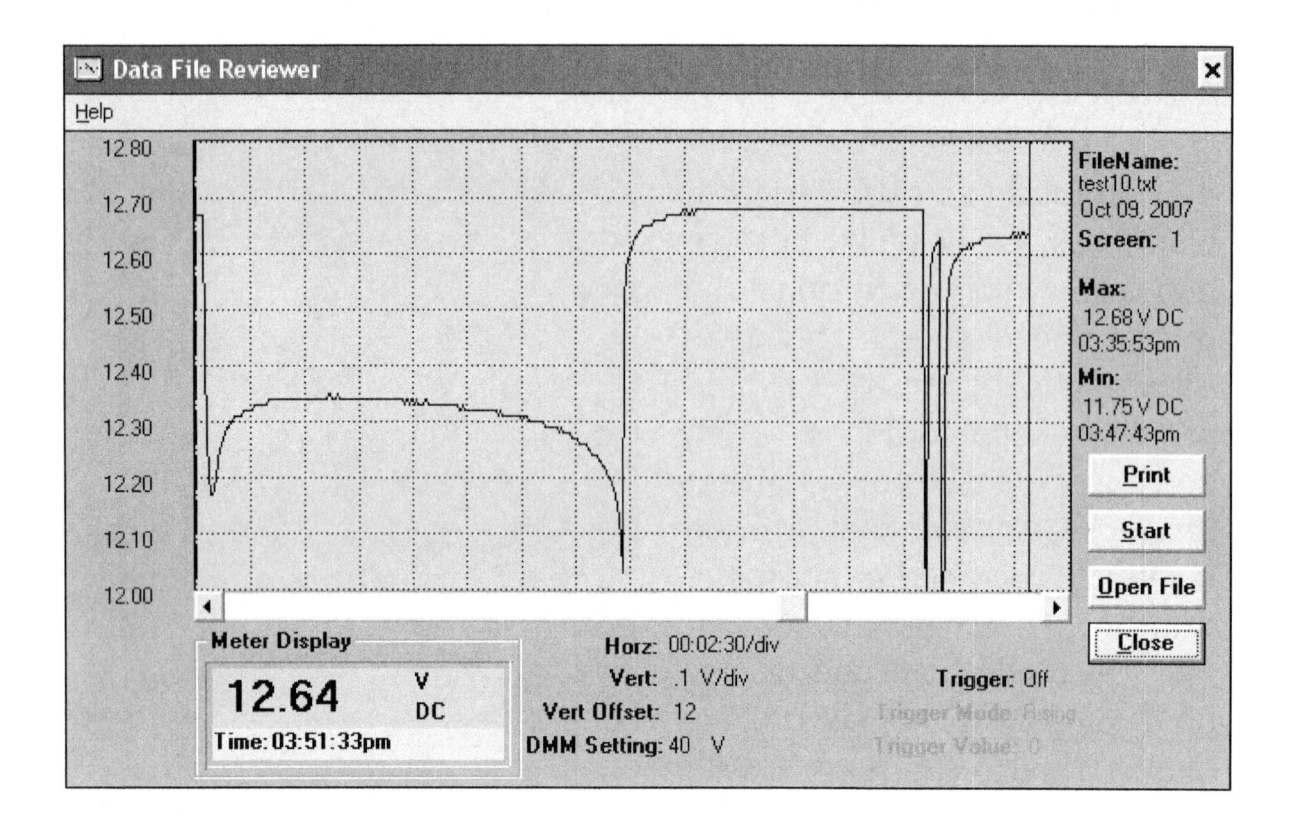

Either method will give good results as long as your log entries are consistent. Logs can lead you to valuable insights and alert you to malfunctions or aberrations during the charging process.

The most useful tests and statistics are:

1. Battery voltage before discharge

2. Battery voltage at end of rest period

3. The time it takes the battery to drop to 12.2 or 11.5 volts during discharge from beginning of discharge.

4. Bounce-back test. This consists of switching off the load when the battery gets to 12.2 or 11.5 volts, waiting 5 minutes, and noting the highest voltage the battery bounces back to and how long it takes to do so.

Wiring and Connectors

5. Spot checks of current draw to battery during charging and current draw to bulb during discharge.

6. Check electrolyte specific gravity after a charge and after a discharge at intervals during formation process. A basic hydrometer as shown in the photograph with temperature integrated thermometer and temperature compensation tables should be used.

Checking specific gravity of the electrolyte

Appendix

Following are lists of resources, and suppliers of materials and tools. All have an internet presence and can be found in search engines with the information given here.

Information Resources

Safety clothing and equipment guidelines

National Institute For Occupational Safety And Health
NIOSH guides and fact sheets are available as free online searchable texts:

NIOSH Pocket Guide To Chemical Hazards

Recommendations For Chemical Protective Clothing

Respirator Fact Sheet

CCOHS (Canadian Center For Occupational Health and Safety)
information about chemical protective clothing and glove selection.

OSHA (Occupational Safety And Health Administration)
Website about lead battery manufacturing.

Department of Transportation

U.S. Environmental Protective Agency

Information Resources

Compliance codes

Many local, state, and federal agencies publish codes or standards on battery compliance solutions. For more information, go to the website **A National Resource of Global Standards** and search by agency or document. The following agencies address stationary lead-acid battery installations:

- **ANSI** — American National Standards Institute
- **ASSE** — American Society of Safety Engineers
- **ASTM** — American Society for Testing Materials
- **CFR** — Code of Federal Regulations
- **IEEE** — Institute of Electrical and Electronics Engineers
- **NFPA** — National Fire Protection Agency
- **OSHA** — Occupational Safety and Health Administration
- **UL** — Underwriters Laboratory Safety equipment and clothing

Professional equipment, safety products and clothing

- **Esca Tech Inc.** — Lead decontamination systems and skin cleaning products
- **McMaster-Carr** — Safety products and clothing
- **New Pig** — Hazmat spill kits
- **DAWG** — Hazmat spill kits
- **The Solar Biz** — Acid spill absorber and neutralizer, absorbent mat
- **Lab Safety Supply** — Industrial and safety supplies
- **Best Gloves** — Safety gloves

Material Data Safety Sheets

MSDS (Material Data Safety Sheets) are available at **Alfa Aesar** for:

- Litharge - Lead (II) oxide, low silver #33330
- Red lead - Lead (II,IV) oxide, red lead oxide, tri-lead tetroxide) #A18066
- CMC (Carboxymethylcellulose sodium salt #A18105

MSDS (H2SO4) specific to Rooto Drain Opener, the Rooto Corporation, (93.2% sulfuric acid) at **Purdue University** MSDS database.

Resin identification codes

If you want to use recycled bottles to store sulfuric acid, you must identify the material the bottle is made of to assess whether or not the material is resistant to sulfuric acid. Please note that even though a bottle may be sulfuric acid resistant, it may not be structurally sound or inadequate for other reasons and thus a safety hazard so you will need to apply to AHJ for final safety compliance.

Resin Identification Code for Bottles at the **Society of the Plastics Industry, Inc.**

DC circuits

Ibiblio www.ibiblio.org/kuphaldt/electricCircuits/

Of interest

Association of Back Yard Metal Casters

Electrochemical Society

Battery Council

Batteries Digest

Materials and Tools

Battery electrolyte

Scholle Chemical Sales

NAPA Auto Parts (also most automotive supply stores)

Sulfuric acid

Alfa Aesar

Photographers Formulary

Rooto Corporation — sulfuric acid drain opener, local hardware or plumbers store

Scientific & Chemicals Supplies Ltd.

Hi-Valley Chemical, Inc.

VWR International

VWR Labshop

Fisher Scientific

Science Kit And Boreal Laboratories

Lead suppliers, general metal suppliers

Mayco Industries

McMaster-Carr

Alchemy Castings

Turkey Foot Trading Company

Rotometals

Nuclead

Online Metals

Metals Depot

Lead oxide (litharge), lead tetraoxide (red lead) and CMC (carboxymethylcellulose)

Scientific & Chemical Supplies Ltd.

Alfa Aesar

Advanced Scientific And Chemical Inc.

Armadillo Clay and Supplies

Sheffield Pottery

Cheap Chemicals.com

Skylighter Chemicals

VWR International

VWR Labshop

Fisher Scientific

Science Kit and Boreal Laboratories

Polypropylene Felt

McMaster-Carr

Hydrometers

Novatech-USA

> Bellweather specific gravity hydrometer 1.000 to 2.000 (subdivisions 0.010) calibrated for 60°F, no thermometer (or with thermometer), length 12.2" #6603-7

> Hydrometer measurement cylinder 500 ml PMP (Polymethylpentene) OD 2", height 15.4" #6230-0500

Ertco – Specific gravity hydrometers

New England Solar – General battery hydrometer with thermometer

Materials and Tools

Plastic Suppliers, sheets, rods, bottles, containers.

McMaster-Carr

United States Plastic Corporation

Vermiculite impregnated fiberglass and carbon fiber welding cloth

McMaster-Carr Cast iron propane stove and LP regulator and hose kit

Northern Tool

Goss Inc.

Bernzomatic ST900D mini torch

Micro-Mark

Bernzomatic JTH-7 hand torch

Hardware store, home supply store

General laboratory items, glassware, 10" & 12" stirring rods

On Line Science Mall

Fanout lead knife

There are many sources for these knives that can be found on the internet, including the following. You should also get a knife sharpening stone, available at any hardware store.

Lincoln Glass Distributors

Maple City Stained Glass

Battery cases, jars, covers

Tulip Products

Battery posts and terminals

McMaster-Carr

Battery post molds (for repairing broken battery posts)
Trimac Molding

Battery post cutter, dry charge battery filler, and hot melt sticks for battery case mending
AW Direct
E-Z Red

PVC syringe, hollow post drill
Battery Equipment Supply

Lead battery post shims
JC Whitney
Solar Seller

Parts for burning rack
McMaster-Carr
Micro-Mark
Jameco

Battery chargers and rectifiers
General automotive stores
Ramsey Electronics
A&A Engineering
David Riddle Co.
CirKits

Knife switches
Filnor Inc.

Materials and Tools

Miscellaneous

DC light bulbs, PV charge controllers and kits, battery anticorrosion rings, wire and cable, battery filler bulbs, power vents, high current diodes, zip cord, heat sinks, fuse, switches.

- Surplus Sales Of Nebraska
- The RF Connection
- CirKits
- Backwoods Solar
- Kansas Wind Power
- New England Solar Electric

Other Titles of Interest

Build Your Own Solar Panel
by Phillip Hurley

Whether you're trying to get off the grid, or you just like to experiment, *Build Your Own Solar Panel* has all the information you need to build your own photovoltaic panel to generate electricity from the sun. The new revised and expanded edition has easy-to-follow directions, and over 150 detailed photos and illustrations. Materials and tools lists, and links to suppliers of PV cells are included. Every-day tools are all that you need to complete these projects.

Build Your Own Solar Panel will show you how to:

- Design and build PV panels
- Customize panel output
- Make tab and bus ribbon
- Solder cell connections
- Wire a photovoltaic panel
- Purchase solar cells
- Test and rate PV cells
- Repair damaged solar cells
- Work with broken cells
- Encapsulate solar cells

Available in print from Amazon.com and for download in full color PDF ebook format at

www.buildasolarpanel.com

Download a free sample of Build Your Own Solar Panel in full color PDF at www.buildasolarpanel.com.

Other Titles of Interest

Solar II:
How to Design, Build and Set Up Photovoltaic Components and Solar Electric Systems
by Phillip Hurley

Now that you've built your solar panels, how do you set up a PV system and plug in? In the e-book Solar II, Phillip Hurley, author of *Build Your Own Solar Panel*, will show you how to:

- Plan and size your solar electric system
- Build racks and charge controllers
- Mount and orient PV panels
- Wire solar panel arrays
- Make a ventilated battery box
- Wire battery arrays for solar panels
- Install an inverter
- Maintain solar batteries for optimum life and performance
- Make your own combiner box, bus bars and DC service box

Solar II includes over 150 photos and illustrations, and a daily power usage calculator. Published April 2007.

Available in print from Amazon.com and for download in full color PDF ebook format at:

www.buildasolarpanel.com

Download a free sample of Solar II in full color PDF at www.buildasolarpanel.com.

Other Titles of Interest

Titles from

Wheelock Mountain Publications:

Other titles by Phillip Hurley:

- Solar II
- Build Your Own Solar Panel
- Build a Solar Hydrogen Fuel Cell System
- Practical Hydrogen Systems
- Build Your Own Fuel Cells
- The Battery Builder's Guide
- Solar Supercapacitor Applications

and also:

- Solar Hydrogen Chronicles *edited by Walt Pyle*
- Tesla: the Lost Inventions *by George Trinkaus*
- Tesla Coil *by George Trinkaus*
- Radio Tesla *by George Trinkaus*
- Son of Tesla Coil *by George Trinkaus*

www.buildasolarpanel.com

Wheelock Mountain Publications
is an imprint of

Good Idea Creative Services
324 Minister Hill Road
Wheelock VT 05851
USA

Printed in Great Britain
by Amazon